ENVIRONMENTAL POLITICS CASEBOOK

Genetically Modified Foods

ENVIRONMENTAL POLITICS CASEBOOK

Genetically Modified Foods

EDITED BY
NORMAN MILLER

LEWIS PUBLISHERS

A CRC Press Company
Boca Raton London New York Washington, D.C.

Library of Congress Cataloging-in-Publication Data

Miller, Norman, 1939-
Environmental politics casebook : genetically modified foods / Norman Miller.
p. cm.
ISBN 1-56670-551-7 (alk. paper)
1. Genetically modified foods—Political aspects—United States. I. Title.

TP248.65.F66 M55 2001
363.19′2—dc21 2001038279

Direct all inquiries to CRC Press LLC, 2000 N.W. Corporate Blvd., Boca Raton, Florida 33431.

Visit the CRC Press Web site at www.crcpress.com

Lewis Publishers is an imprint of CRC Press LLC

International Standard Book Number 1-56670-551-7
Library of Congress Card Number 2001038279
Printed in the United States of America 1 2 3 4 5 6 7 8 9 0
Printed on acid-free paper

Introduction

This *Casebook* companion to *Environmental Politics: Interest Groups, the Media, and the Making of Policy* is a compilation of the documents that, collectively, constitute the political profile of a major environmental issue — the bioengineering of food. The thesis of *Environmental Politics* is that public policy derives from the reconciliation of the competing interests with stakes in how an issue is resolved. But the competition among these interests does not take place in the abstract; rather it is embodied in the legislation and regulations, media stories and op-ed pieces, speeches and Web sites, international agreements, and scientific papers that the various interests prepare and advance to influence and shape that policy. In assembling and organizing these materials I have attempted to illustrate and exemplify the concepts and principles in the text, to put flesh on the bones of its generalizations and assertions.

I have also provided brief introductions to the sections of the *Casebook*, which correspond to the chapters in the text. My purpose in doing so is to encourage students to follow their study and discussion of a text chapter with a reading of the corresponding primary sources in the *Casebook*, again with a view toward reinforcing the principles in the text with real-life examples. These introductions also include the identification of some of the questions and philosophical issues they raise and, where appropriate, suggestions for further reading. Hopefully they will also generate discussion and constructive controversy. When the student finishes both volumes, he or she will have a stronger conceptual grasp of how public policies are developed and a concrete sense of how they play out in an actual, present situation.

The issue selected for this case study is genetically modified food, variously called in the literature "genetically altered food," "bioengineered food," or, colloquially, GM food, or GMOs, for the organisms responsible for the "modification." As I reviewed the many environmental problems and concerns that had generated legislative and media attention of late, the issue of the genetic alteration of food stood out. For several years, it has been one of the most hotly debated practices, and hardly a week goes by without some new journal article, scientific claim, op-ed piece, or newspaper advertisement taking it up. In many respects, it is a paradigm of the kind of issue that is capturing legislative agendas, the kind that touches most of the population in an intimate way in their daily lives. Being of relatively recent origin, genetically modified foods have not been systematically tested over a long period of time. The scientific uncertainty inherent in this situation makes it all the more attractive to the media, which have commercially exploited the potentially serious and insidious health effects on all of us who eat them. The conflict between those who regard GMOs as another example of science gone mad and those to whom they are an important piece of the world hunger solution is of sufficient magnitude to warrant dramatic treatment. And the active participation of international actors and cultural forces has not only proven a catalyst for political unrest here in the U.S., but threatens the economic viability of domestic agriculture and a segment of the biotech industry itself.

Finally, if some strange transgenic organism should somehow escape into a field of vulnerable vegetation, the ecosystem itself could somehow be altered in irremediable ways. In short, this issue represents a rich mine of controversy, and, hence, a treasure trove for those who would seek to understand how the interplay of interest group forces leads to policy.

Table of Contents

Section I

Legislative Documents

Included in this section are federal bills H.R. 3377 and H.R. 3883 (106th Congress, 1st and 2nd st Sessions, respectively), both sponsored by Representative Dennis Kucinich, and short-titled, respectively, the "Genetically Engineered Food Right to Know Act" and the "Genetically Engineered Food Safety Act." In the Senate, Senator Barbara Boxer introduced S. 2080, also titled "The Genetically Engineered Food Right to Know Act," which is almost identical to Congressman Kucinich's, but which includes an appropriation of $5 million to make grants to study the possible health and environmental risks associated with genetically engineered foods. Senator Patrick Moynihan introduced S. 2315, the "Genetically Engineered Food Safety Act," in the Senate as well.

H.R. 3377 and S. 2080, as their titles suggest, would require producers and manufacturers to label all foods that contain, or are produced with, genetically engineered material, clearly prescribing the form of the label and the subtext, *UNITED STATES GOVERNMENT NOTICE: THIS PRODUCT CONTAINS GENETICALLY ENGINEERED MATERIAL, OR WAS PRODUCED WITH GENETICALLY ENGINEERED MATERIAL.* Their respective "Findings" sections justify the proposed measures by pointing to the as yet unknown risks that may be posed by genetically engineered foods and the right of the public to know if the foods they purchase and consume contain, or are produced with, genetically modified material. Also included after the text of H.R. 3377 is a more extensive statement by Senator Boxer, elaborating on her legislation and the need for it.

Mentioned in supporting statements, though not in the bills themselves, are potential *ecological* risks. Widely publicized experiments at New York's Cornell University found potentially adverse effects on Monarch butterflies who were fed genetically engineered food. (See Section 8 on Science.)

Two other things should be noted. First, H.R. 3377 had more than 18 joint sponsors and over 50 co-sponsors, crossing partisan and geographical lines. Second, the House bill was referred to both the Agriculture and Commerce Committees, reflecting the breadth of its potential impact.

H.R. 3883 was introduced by Congressman Kucinich approximately 4 months after his labeling bill. Included is an extensive statement, published in the Congressional Record in May of 1999 by him and Congresswoman Kaptur, justifying the initiative. In May, Senator Moynihan, with co-sponsors Senators Reid and Boxer, introduced their bill, which is substantively identical to Representative Kucinich's. H.R. 3883 and S. 2315 bring "genetic engineering" under the umbrella of "food additives," and thus subject food produced from, or involving, genetic materials to the safety testing protocol required under the federal "Food, Drug, and Cosmetic Act."

Discussion

H.R. 3377 and S. 2080 are among the latest in a series of measures that represent a new direction in regulation — mandatory disclosure — or what a recent analyst characterized as "regulation by shaming." Like predecessor mandates to disclose, e.g., toxic emissions, public water supply contaminants, automobile gas mileage performance, nutritional breakdowns of processed foods, even the methods by which fish sold in cans are caught (witness the "dolphin-free" tuna notices on their packages), these bills ostensibly go toward the unimpeachable goal of informed consumer choice.

It might be useful, first, to discuss the advantages and limitations of "regulation by disclosure" for both the producer/manufacturer and the consumer. It then might be instructive to apply the conclusions reached to this particular situation. Unlike previous disclosure requirements, what is to be disclosed under these bills is of as yet unknown scientific or health consequence. While no one at this point can definitively affirm the safety of genetically derived food, neither is there any scientific evidence of any adverse health effects on humans. Thus, it has been argued that consumers cannot constructively act upon the information that is required to be disclosed, except to the extent that they can avoid the unknown. This, too, has positives and negatives. The concept noted above, "regulation by shaming," is discussed in provocative detail in an article by the same name, written by Mary Graham, which appears in the April 2000 issue of the *Atlantic*. It is a good starting point for a discussion of this issue.

The conclusion that the information required to be disclosed in this case has no immediate practical application might lead one to suspect that there are other motives for advancing the measures. After all, we in this country have been consuming genetically engineered food in one form or another for several years without any apparent ill effects, and there have been no allegations to the contrary. And a single study such as the one at Cornell suggesting potential ecological threat is too narrow in its application and too tentative to warrant the level of Congressional attention the issue has garnered. There are many "canaries in coal mines" out there that are routinely ignored or dismissed by our lawmakers for this to be taken at face value. Further, the European Union established labeling requirements 2 years earlier, and no similar action by the U.S. Congress was even discussed.

Given the timing of this legislation, and the growing restiveness abroad about genetically modified foods, the driving factor behind this legislation could be seen as the protection of the American agricultural community and its exports rather than the protection of the American consumer. See the sections on the international reaction to genetically modified food, and on how the media abroad has publicized it to American citizens, to put this into context. Note also that the bill was referred to the Commerce Committee as well as the Agriculture Committee.

Initial co-sponsors of these bills were predominantly, though not exclusively, Democrats, and the organizations supporting them were generally pro-environment and/or pro-consumer. When, if ever, do Republicans come on board? Does Democratic sponsorship doom consideration of the measures in Committees and, ultimately, their enactment, given the prevailing Congressional Republican majorities?

Finally, H.R. 3883 and S. 2315 mandating research into the effects of these new technologically implicated food products were introduced after, rather than before, the bills mandating labeling. Unless the legislative intent is for the raised public consciousness and unease over GM food to increase public pressure for passage, the normal course of action would be to identify the problem before taking steps to address it. Since public confidence in the research conducted by our regulatory agencies in this case is marginal, is extending the protocol for food safety to GM foods anything more than a token measure?

It would also be interesting to study the differences, if any, between these bills and their anticipated counterparts in the succeeding Congressional session, where we have a Republican rather than a Democratic president and different partisan breakdowns in both Houses. Comparing committee references and co-sponsors might also prove enlightening.

106TH CONGRESS
1ST SESSION

H. R. 3377

To amend the Federal Food, Drug, and Cosmetic Act, the Federal Meat Inspection Act, and the Poultry Products Inspection Act to require that food that contains a genetically engineered material, or that is produced with a genetically engineered material, be labeled accordingly.

IN THE HOUSE OF REPRESENTATIVES

NOVEMBER 16, 1999

Mr. KUCINICH (for himself, Mr. METCALF, Mr. BONIOR, Mr. DEFAZIO, Mr. SANDERS, Mr. SMITH of New Jersey, Mr. DOYLE, Mr. LIPINSKI, Mr. BROWN of Ohio, Mr. HINCHEY, Ms. SCHAKOWSKY, Ms. NORTON, Mr. STARK, Ms. WOOLSEY, Mrs. MINK of Hawaii, Mr. MARTINEZ, Mr. MCDERMOTT, Ms. LEE, and Ms. WATERS) introduced the following bill; which was referred to the Committee on Agriculture, and in addition to the Committee on Commerce, for a period to be subsequently determined by the Speaker, in each case for consideration of such provisions as fall within the jurisdiction of the committee concerned

---

A BILL

To amend the Federal Food, Drug, and Cosmetic Act, the Federal Meat Inspection Act, and the Poultry Products Inspection Act to require that food that contains a genetically engineered material, or that is produced with a genetically engineered material, be labeled accordingly.

Be it enacted by the Senate and House of Representatives of the United States of America in Congress assembled,

SECTION 1. SHORT TITLE; TABLE OF CONTENTS.

(a) SHORT TITLE.— This Act may be cited as the "Genetically Engineered Food Right to Know Act".

(b) TABLE OF CONTENTS.— The table of contents of this Act is as follows:

SEC. 2. FINDINGS.

The Congress finds as follows:

(1) The process of genetically engineering foods results in the material change of such foods.

(2) The Congress has previously required that all foods bear labels that reveal material facts to consumers.

(3) Federal agencies have failed to uphold Congressional intent by allowing genetically engineered foods to be marketed, sold and otherwise used without labeling that reveals material facts to the public.

(4) Consumers wish to know whether the food they purchase and consume contains or is produced with a genetically engineered material for a variety of reasons, including the potential transfer of allergens into food and other health risks, concerns about potential environmental risks associated with the genetic engineering of crops, and religiously and ethically based dietary restrictions.

(5) Consumers have a right to know whether the food they purchase contains or was produced with genetically engineered material.

(6) Reasonably available technology permits the detection in food of genetically engineered material, generally acknowledged to be as low as 0.1 percent.

SEC. 3. LABELING REGARDING GENETICALLY ENGINEERED MATERIAL; AMENDMENTS TO FEDERAL FOOD, DRUG, AND COSMETIC ACT.

(a) IN GENERAL.— Section 403 of the Federal Food, Drug, and Cosmetic Act (21 U.S.C. 343) is amended by adding at the end the following paragraph:

"(t)(1) If it contains a genetically engineered material, or was produced with a genetically engineered material, unless it bears a label (or labeling, in the case of a raw agricultural commodity, other than the sale of such a commodity at retail) that provides notices in accordance with the following:

"(A) A notice as follows: 'GENETICALLY ENGINEERED'.

"(B) A notice as follows: 'UNITED STATES GOVERNMENT NOTICE: THIS PRODUCT CONTAINS A GENETICALLY ENGINEERED MATERIAL, OR WAS PRODUCED WITH A GENETICALLY ENGINEERED MATERIAL'.

"(C) The notice required in clause (A) immediately precedes the notice required in clause (B) and is not less than twice the size of the notice required in clause (B).

"(D) The notice required in clause (B) is of the same size as would apply if the notice provided nutrition information that is required in paragraph (q)(1).

"(E) The notices required in clauses (A) and (B) are clearly legible and conspicuous.

"(2) For purposes of subparagraph (1):

"(A) The term 'genetically engineered material' means material derived from any part of a genetically engineered organism, without regard to whether the altered molecular or cellular characteristics of the organism are detectable in the material.

"(B) The term 'genetically engineered organism' means—

"(i) an organism that has been altered at the molecular or cellular level by means that are not possible under natural conditions or processes (including but not limited to recombinant DNA and RNA techniques, cell fusion, microencapsulation, macroencapsulation, gene deletion and doubling, introducing a foreign gene, and changing the positions of genes), other than a means consisting exclusively of breeding, conjugation, fermentation, hybridization, in vitro fertilization, or tissue culture, and

"(ii) an organism made through sexual or asexual reproduction (or both) involving an organism described in subclause (i), if possessing any of the altered molecular or cellular characteristics of the organism so described.

"(3) For purposes of subparagraph (1), a food shall be considered to have been produced with a genetically engineered material if—

"(A) the organism from which the food is derived has been injected or otherwise treated with a genetically engineered material (except that the use of manure as a fertilizer for raw agricultural commodities may not be construed to mean that such commodities are produced with a genetically engineered material);

"(B) the animal from which the food is derived has been fed genetically engineered material, or

"(C) the food contains an ingredient that is a food to which clause (A) or (B) applies.

"(4) This paragraph does not apply to food that—

"(A) is served in restaurants or other establishments in which food is served for immediate human consumption,

"(B) is processed and prepared primarily in a retail establishment, is ready for human consumption, which is of the type described in clause (A), and is offered for sale to consumers but not for immediate human consumption in such establishment and is not offered for sale outside such establishment, or

"(C) is a medical food as defined in section 5(b) of the Orphan Drug Act.".

(b) CIVIL PENALTIES.— Section 303 of the Federal Food, Drug, and Cosmetic Act (21 U.S.C. 333) is amended by adding at the end the following subsection:

"(h)(1) With respect to a violation of section 301(a), 301(b), or 301(c) involving the misbranding of food within the meaning of section 403(t), any person engaging in such a violation shall be liable to the United States for a civil penalty in an amount not to exceed $100,000 for each such violation.

"(2) Paragraphs (3) through (5) of subsection (g) apply with respect to a civil penalty under paragraph (1) of this subsection to the same extent and in the same manner as such paragraphs (3) through (5) apply with respect to a civil penalty under paragraph (1) or (2) of subsection (g).".

(c) GUARANTY.—

(1) IN GENERAL.— Section 303(d) of the Federal Food, Drug, and Cosmetic Act (21 U.S.C. 333(d)) is amended—

(A) by striking "(d)" and inserting "(d)(1)"; and

(B) by adding at the end the following paragraph:

"(2)(A) No person shall be subject to the penalties of subsection (a)(1) or (h) for a violation of section 301(a), 301(b), or 301(c) involving the misbranding of food within the meaning of section 403(t) if such person (referred to in this paragraph as the 'recipient') establishes a guaranty or undertaking signed by, and containing the name and address of, the person residing in the United States from whom the recipient received in good faith the food (including the receipt of seeds to grow raw agricultural commodities), to the effect that (within the meaning of section 403(t)) the food does not contain a genetically engineered material or was not produced with a genetically engineered material.

"(B) In the case of a recipient who with respect to a food establishes a guaranty or undertaking in accordance with subparagraph (A), the exclusion under such subparagraph from being subject to penalties applies to the recipient without regard to the use of the food by the recipient, including—

"(i) processing the food,

"(ii) using the food as an ingredient in a food product,

"(iii) repacking the food, or

"(iv) growing, raising, or otherwise producing the food.".

(2) FALSE GUARANTY.— Section 301(h) of the Federal Food, Drug, and Cosmetic Act (21 U.S.C. 331(h)) is amended by inserting "or 303(d)(2)" after "303(c)(2)".

(d) UNINTENDED CONTAMINATION.— Section 303(d) of the Federal Food, Drug, and Cosmetic Act, as amended by subsection (c)(1) of this section, is amended by adding at the end the following paragraph:

"(3)(A) No person shall be subject to the penalties of subsection (a)(1) or (h) for a violation of section 301(a), 301(b), or 301(c) involving the misbranding of food within the meaning of section 403(t) if—

"(i) such person is an agricultural producer and the violation occurs because food that is grown, raised, or otherwise produced by such producer, which food does not contain a genetically engineered material and was not produced with a genetically engineered material, is contaminated with a food that contains a genetically engineered material or was produced with a genetically engineered material (including contamination by mingling the two), and

"(ii) such contamination is not intended by the agricultural producer.

"(B) Subparagraph (A) does not apply to an agricultural producer to the extent that the contamination occurs as a result of the negligence of the producer.".

SEC. 4. LABELING REGARDING GENETICALLY ENGINEERED MATERIAL; AMENDMENTS TO FEDERAL MEAT INSPECTION ACT.

(a) REQUIREMENTS.— The Federal Meat Inspection Act is amended by inserting after section 7 (21 U.S.C. 607) the following section:

"SEC. 7A. REQUIREMENTS FOR LABELING REGARDING GENETICALLY ENGINEERED MATERIAL.

"(a) DEFINITIONS.— In this section:

"(1) The term 'meat food' means a carcass, part of a carcass, meat, or meat food product that is derived from cattle, sheep, swine, goats, horses, mules, or other equines and is capable of use as human food.

"(2) The term 'genetically engineered material' means material derived from any part of a genetically engineered organism, without regard to whether the altered molecular or cellular characteristics of the organism are detectable in the material (and without regard to whether the organism is capable of use as human food).

"(3) The term 'genetically engineered organism' means—

"(A) an organism that has been altered at the molecular or cellular level by means that are not possible under natural conditions or processes (including but not limited to recombinant DNA and RNA techniques, cell fusion, microencapsulation, macroencapsulation, gene deletion and doubling, introducing a foreign gene, and changing the positions of genes), other than a means consisting exclusively of breeding, conjugation, fermentation, hybridization, in vitro fertilization, or tissue culture; and

"(B) an organism made through sexual or asexual reproduction (or both) involving an organism described in subparagraph (A), if possessing any of the altered molecular or cellular characteristics of the organism so described.

"(b) LABELING REQUIREMENT.—

"(1) REQUIRED LABELING TO AVOID MISBRANDING.— For purposes of sections 1(n) and 10, a meat food is misbranded if it—

"(A) contains a genetically engineered material or was produced with a genetically engineered material; and

"(B) does not bear a label (or include labeling, in the case of a meat food that is not packaged in a container) that provides, in a clearly legible and conspicuous manner, the notices described in subsection (c).

"(2) RULE OF CONSTRUCTION.— For purposes of paragraph (1)(A), a meat food shall be considered to have been produced with a genetically engineered material if—

"(A) the organism from which the food is derived has been injected or otherwise treated with a genetically engineered material;

"(B) the animal from which the food is derived has been fed genetically engineered material; or

"(C) the food contains an ingredient that is a food to which subparagraph (A) or (B) applies.

"(c) SPECIFICS OF LABEL NOTICES.—

"(1) REQUIRED NOTICES.— The notices referred to in subsection (b)(1)(B) are the following:

"(A) A notice as follows: 'GENETICALLY ENGINEERED'.

"(B) A notice as follows: 'UNITED STATES GOVERNMENT NOTICE: THIS PRODUCT CONTAINS A GENETICALLY ENGINEERED MATERIAL, OR WAS PRODUCED WITH A GENETICALLY ENGINEERED MATERIAL'.

"(2) LOCATION AND SIZE.— (A) The notice required in paragraph (1)(A) shall immediately precede the notice required in paragraph (1)(B) and shall be not less than twice the size of the notice required in paragraph (1)(B).

"(B) The notice required in paragraph (1)(B) shall be of the same size as would apply if the notice provided nutrition information that is required in section 403(q)(1) of the Federal Food, Drug, and Cosmetic Act.

"(d) EXCEPTIONS TO REQUIREMENTS.— Subsection (a) does not apply to any meat food that—

"(1) is served in restaurants or other establishments in which food is served for immediate human consumption; or

"(2) is processed and prepared primarily in a retail establishment, is ready for human consumption, is offered for sale to consumers but not for immediate human consumption in such establishment, and is not offered for sale outside such establishment.

"(e) GUARANTY.—

"(1) IN GENERAL.— A packer, processor, or other person shall not be considered to have violated the requirements of this section with respect to the labeling of meat food if the packer, processor, or other person (referred to in this subsection as the 'recipient') establishes a guaranty or undertaking signed by, and containing the name and address of, the person residing in the United States from whom the recipient received in good faith the meat food or the animal from which the meat food was derived, or received in good faith food intended to be fed to such animal, to the effect that the meat food, or such animal, or such food, respectively, does not contain genetically engineered material or was not produced with a genetically engineered material.

"(2) SCOPE OF GUARANTY.— In the case of a recipient who establishes a guaranty or undertaking in accordance with paragraph (1), the exclusion under such paragraph from being subject to penalties applies to the recipient without regard to the use of the meat food by the recipient (or the use by the recipient of the animal from which the meat food was derived, or of food intended to be fed to such animal), including—

"(A) processing the meat food;

"(B) using the meat food as an ingredient in another food product;

"(C) packing or repacking the meat food; or

"(D) raising the animal from which the meat food was derived.

"(3) FALSE GUARANTY.— It is a violation of this Act for a person to give a guaranty or undertaking in accordance with paragraph (1) that the person knows or has reason to know is false.

"(f) CIVIL PENALTIES.—

"(1) IN GENERAL.— The Secretary may assess a civil penalty against a person that violates subsection (b) or (c)(3) in an amount not to exceed $100,000 for each such violation.

"(2) NOTICE AND OPPORTUNITY FOR HEARING.— A civil penalty under paragraph (1) shall be assessed by the Secretary by an order made on the record after opportunity for a hearing provided in accordance with this subparagraph and section 554 of title 5, United States Code. Before issuing such an order, the Secretary shall give written notice to the person to be assessed a civil penalty under such order of the Secretary's proposal to issue such order and provide such person an opportunity for a hearing on the order. In the course of any investigation, the Secretary may issue subpoenas requiring the attendance and testimony of witnesses and the production of evidence that relates to the matter under investigation.

"(3) CONSIDERATIONS REGARDING AMOUNT OF PENALTY.— In determining the amount of a civil penalty under paragraph (1), the Secretary shall take into account the nature, circumstances, extent, and gravity of the violation or violations and, with respect to the violator, ability to pay, effect on ability to continue to do business, any history of prior such violations, the degree of culpability, and such other matters as justice may require.

"(4) CERTAIN AUTHORITIES.— The Secretary may compromise, modify, or remit, with or without conditions, any civil penalty under paragraph (1). The amount of such penalty, when finally determined, or the amount agreed upon in compromise, may be deducted from any sums owing by the United States to the person charged.

"(5) JUDICIAL REVIEW.— Any person who requested, in accordance with paragraph (2), a hearing respecting the assessment of a civil penalty under paragraph (1) and who is aggrieved by an order assessing a civil penalty may file a petition for judicial review of such order with the United States Court of Appeals for the District of Columbia Circuit or for any other circuit in which such person resides or transacts business. Such a petition may only be filed within the 60-day period beginning on the date the order making such assessment was issued.

"(6) FAILURE TO PAY.— If a person fails to pay an assessment of a civil penalty—

"(A) after the order making the assessment becomes final, and if such person does not file a petition for judicial review of the order in accordance with paragraph (5); or

"(B) after a court in an action brought under paragraph (4) has entered a final judgment in favor of the Secretary;

the Attorney General shall recover the amount assessed (plus interest at currently prevailing rates from the date of the expiration of the 60-day period referred to in paragraph (5) or the date of such final judgment, as the case may be) in an action brought in any appropriate district court of the United States. In such an action, the validity, amount, and appropriateness of such penalty shall not be subject to review.".

(b) INCLUSION OF LABELING REQUIREMENTS IN DEFINITION OF MISBRANDED.— Section 1(n) of the Federal Meat Inspection Act (21 U.S.C. 601(n)) is amended—

(1) by striking "or" at the end of paragraph (11);

(2) by striking the period at the end of paragraph (12) and inserting "; or"; and

(3) by adding at the end the following paragraph:

"(13) if it fails to bear a label or labeling as required by section 7A.".

SEC. 5. LABELING REGARDING GENETICALLY ENGINEERED MATERIAL; AMENDMENTS TO POULTRY PRODUCTS INSPECTION ACT.

The Poultry Products Inspection Act is amended by inserting after section 8 (21 U.S.C. 457) the following section:

"SEC. 8A. REQUIREMENTS FOR LABELING REGARDING GENETICALLY ENGINEERED MATERIAL.

"(a) DEFINITIONS.— In this section:

"(1) The term 'genetically engineered material' means material derived from any part of a genetically engineered organism, without regard to whether the altered molecular or cellular characteristics of the organism are detectable in the material (and without regard to whether the organism is capable of use as human food).

"(2) The term 'genetically engineered organism' means—

"(A) an organism that has been altered at the molecular or cellular level by means that are not possible under natural conditions or processes (including but not limited to recombinant DNA and RNA techniques, cell fusion, microencapsulation, macroencapsulation, gene deletion and doubling, introducing a foreign gene, and changing the positions of genes), other than a means consisting exclusively of breeding, conjugation, fermentation, hybridization, in vitro fertilization, or tissue culture; and

"(B) an organism made through sexual or asexual reproduction (or both) involving an organism described in subparagraph (A), if possessing any of the altered molecular or cellular characteristics of the organism so described.

"(b) LABELING REQUIREMENT.—

"(1) REQUIRED LABELING TO AVOID MISBRANDING.— For purposes of sections 4(h) and 9(a), a poultry product is misbranded if it—

"(A) contains a genetically engineered material or was produced with a genetically engineered material; and

"(B) does not bear a label (or include labeling, in the case of a poultry product that is not packaged in a container) that provides, in a clearly legible and conspicuous manner, the notices described in subsection (c).

"(2) RULE OF CONSTRUCTION.— For purposes of paragraph (1)(A), a poultry product shall be considered to have been produced with a genetically engineered material if—

"(A) the poultry from which the food is derived has been injected or otherwise treated with a genetically engineered material;

"(B) the poultry from which the food is derived has been fed genetically engineered material; or

"(C) the food contains an ingredient that is a food to which subparagraph (A) or (B) applies.

"(c) SPECIFICS OF LABEL NOTICES.—

"(1) REQUIRED NOTICES.— The notices referred to in subsection (b)(1)(B) are the following:

"(A) A notice as follows: 'GENETICALLY ENGINEERED'.

"(B) A notice as follows: 'UNITED STATES GOVERNMENT NOTICE: THIS PRODUCT CONTAINS A GENETICALLY ENGINEERED MATERIAL, OR WAS PRODUCED WITH A GENETICALLY ENGINEERED MATERIAL'.

"(2) LOCATION AND SIZE.— (A) The notice required in paragraph (1)(A) shall immediately precede the notice required in paragraph (1)(B) and shall be not less than twice the size of the notice required in paragraph (1)(B).

"(B) The notice required in paragraph (1)(B) shall be of the same size as would apply if the notice provided nutrition information that is required in section 403(q)(1) of the Federal Food, Drug, and Cosmetic Act.

"(d) EXCEPTIONS TO REQUIREMENTS.— Subsection (a) does not apply to any poultry product that—

"(1) is served in restaurants or other establishments in which food is served for immediate human consumption; or

"(2) is processed and prepared primarily in a retail establishment, is ready for human consumption, is offered for sale to consumers but not for immediate human consumption in such establishment, and is not offered for sale outside such establishment.

"(e) GUARANTY.—

"(1) IN GENERAL.— An official establishment or other person shall not be considered to have violated the requirements of this section with respect to the labeling of a poultry product if the official establishment or other person (referred to in this subsection as the 'recipient') establishes a guaranty or undertaking signed by, and containing the name and address of, the person residing in the United States from whom the recipient received in good faith the poultry product or the poultry from which the poultry product was derived, or received in good faith food intended to be fed to poultry, to the effect that the poultry product, poultry, or such food, respectively, does not contain genetically engineered material or was not produced with a genetically engineered material.

"(2) SCOPE OF GUARANTY.— In the case of a recipient who establishes a guaranty or undertaking in accordance with paragraph (1), the exclusion under such paragraph from being subject to penalties applies to the recipient without regard to the use of the poultry product by the recipient (or the use by the recipient of the poultry from which the poultry product was derived, or of food intended to be fed to such poultry), including—

"(A) processing the poultry;

"(B) using the poultry product as an ingredient in another food product;

"(C) packing or repacking the poultry product; or

"(D) raising the poultry from which the poultry product was derived.

"(3) FALSE GUARANTY.— It is a violation of this Act for a person to give a guaranty or undertaking in accordance with paragraph (1) that the person knows or has reason to know is false.

"(f) CIVIL PENALTIES.—

"(1) IN GENERAL.— The Secretary may assess a civil penalty against a person that violates subsection (b) or (c)(3) in an amount not to exceed $100,000 for each such violation.

"(2) NOTICE AND OPPORTUNITY FOR HEARING.— A civil penalty under paragraph (1) shall be assessed by the Secretary by an order made on the record after opportunity for a hearing provided in accordance with this subparagraph and section 554 of title 5, United States Code. Before issuing such an order, the Secretary shall give written notice to the person to be assessed a civil penalty under such order of the Secretary's proposal to issue such order and provide such person an opportunity for a hearing on the order. In the course of any investigation, the Secretary may issue subpoenas requiring the attendance and testimony of witnesses and the production of evidence that relates to the matter under investigation.

"(3) CONSIDERATIONS REGARDING AMOUNT OF PENALTY.— In determining the amount of a civil penalty under paragraph (1), the Secretary shall take into account the nature, circumstances, extent, and gravity of the violation or violations and, with respect to the violator, ability to pay, effect on ability to continue to do business, any history of prior such violations, the degree of culpability, and such other matters as justice may require.

"(4) CERTAIN AUTHORITIES.— The Secretary may compromise, modify, or remit, with or without conditions, any civil penalty under paragraph (1). The amount of such penalty, when finally determined, or the amount agreed upon in compromise, may be deducted from any sums owing by the United States to the person charged.

"(5) JUDICIAL REVIEW.— Any person who requested, in accordance with paragraph (2), a hearing respecting the assessment of a civil penalty under paragraph (1) and who is aggrieved by an order assessing a civil penalty may file a petition for judicial review of such order with the United States Court of Appeals for the District of Columbia Circuit or for any other circuit in which such person resides or transacts business. Such a petition may only be filed within the 60-day period beginning on the date the order making such assessment was issued.

"(6) FAILURE TO PAY.— If a person fails to pay an assessment of a civil penalty—

"(A) after the order making the assessment becomes final, and if such person does not file a petition for judicial review of the order in accordance with paragraph (5); or

"(B) after a court in an action brought under paragraph (4) has entered a final judgment in favor of the Secretary;

the Attorney General shall recover the amount assessed (plus interest at currently prevailing rates from the date of the expiration of the 60-day period referred to in paragraph (5) or the date of such final judgment, as the case may be) in an action brought in any appropriate district court of the United States. In such an action, the validity, amount, and appropriateness of such penalty shall not be subject to review.".

(b) INCLUSION OF LABELING REQUIREMENTS IN DEFINITION OF MISBRANDED.— Section 4(h) of the Poultry Products Inspection Act (21 U.S.C. 453(h)) is amended—

(1) by striking "or" at the end of paragraph (11);

(2) by striking the period at the end of paragraph (12) and inserting "; or"; and

(3) by adding at the end the following paragraph:

"(13) if it fails to bear a label or labeling as required by section 8A.".

SEC. 6. EFFECTIVE DATE.

This Act and the amendments made by this Act take effect upon the expiration of the 180-day period beginning on the date of the enactment of this Act.

THE GENETICALLY ENGINEERED FOOD RIGHT-TO-KNOW ACT, S. 2080

Mrs. BOXER. Mr. President, today I am pleased to introduce the Genetically Engineered Food Right-to-Know Act. This legislation requires that all foods containing or produced with genetically engineered material bear a neutral label stating that: 'this product contains a genetically engineered material or was produced with a genetically engineered material.'

The bill adds this labeling requirement to the provisions of the Federal Food, Drug, and Cosmetic Act (FFDCA), the Federal Meat Inspection Act, and the Poultry Products Inspection Act which contain the general standards for labeling foods.

Recent polls have demonstrated that Americans want to know if they are eating genetically engineered food. A January 1999 Time magazine poll revealed that 81% of respondents wanted genetically engineered food to be labeled. A January 2000 MSNBC poll showed identical results.

This pressure has already led some companies not to use genetically engineered materials in their foods. Gerber and Heinz have said they will no longer use genetically engineered material in their baby food. Whole Foods and Wild Oats Supermarkets also have said they will use no genetically engineered material in their own products.

Great Britain, France, Germany, the Netherlands, Belgium, Luxembourg, Denmark, Sweden, Finland, Ireland, Spain, Austria, Italy, Portugal, Greece, New Zealand, and Japan already require genetically engineered food to be labeled.

If the U.S. wants to sell its genetically engineered food to these countries, it will have to label the food for foreign consumers. It is only fair that American consumers be given similar information.

Why do I feel it's important for consumers to know that their food is genetically engineered?

First, we don't know whether genetically engineered food is harmful or whether it is safe. However, scientists have raised concerns about genetically engineered food. These concerns include the risks of increased exposure to allergens, decreased nutritional value, increased toxicity and increased antibiotic resistance.

In addition, scientists have raised concerns about the ecological risks associated with genetically engineered food. Some of those risks include the destruction of species, cross pollination that breeds new weeds that are resistant to herbicides, and increases in pesticide use over the long-term.

Earlier this year, for example, researchers at Cornell University reported that Monarch butterflies were either killed or developed abnormally when eating milkweed dusted with the pollen of Bt-corn, a genetically engineered food.

Second, the Food and Drug Administration does not require pre-market health and safety testing of genetically engineered foods. Therefore, it is only fair that consumers know they are eating products that have not been tested.

Third, the Environmental Protection Agency and the Department of Agriculture do not require substantive environmental review of genetically engineered materials under their jurisdiction.

My Genetically Engineered Food Right-to-Know Act not only mandates labels, but does something even more important: it authorizes $5 million in grants to conduct studies into the health and environmental risks raised by genetically engineered food.

Specifically, it directs the Secretary of HHS to make grants to individuals, organizations and institutions to study risks like increased toxicity, increased allergenicity, negative effects on soil ecology and on the environment in general.

What is the extent of genetically engineered crops today?

Last year, 98.6 million acres in the U.S. were planted with genetically engineered crops. More than one-third of the U.S. soybean crop and one-quarter of corn were genetically engineered. This represents a 23-fold increase in genetically engineered crop production from just four years ago.

And waiting to come into the marketplace are more than 60 different genetically engineered crops--from apples and strawberries to potatoes and tomatoes.

Providing consumers with information about the foods they eat is hardly new.

For example, I was proud to be the author of the law to provide for the 'dolphin safe' label on tuna. The label indicated that the tuna was harvested by methods that don't harm dolphins.

I was also proud to lead the fight in the Senate to make sure that chicken frozen as solid as a blowing ball could not be labeled fresh. At the time, USDA's position was that frozen chicken could be labeled 'fresh.'

In 1996, I succeeded in amending the Safe Drinking Water Act to require that drinking water providers give their consumers annual reports concerning the quality of their water.

Others in Congress led the fight to tell consumers whether their products contain artificial colors or sweeteners, preservatives, additives, and whether they are from concentrate. I supported those labels as well.

Food manufacturers also label their products with information that is of little value to consumers. Certain brands of pretzels, for example, bear a label which states that the manufacturer is a 'Member of the Snack Food Association: An International Trade Association.'

I don't think this is information consumers are clamoring for, yet the manufacturer is willing to go through the trouble of putting it on the bag.

My legislation builds on the existing food labeling system, and would be simple to implement. It would require that all foods containing or made with genetically engineered foods be labeled with this information: 'this product contains a genetically engineered material or was produced with a genetically engineered material.'

For example, corn flakes made with genetically engineered corn would be a 'product that contains' genetically engineered material. To take another example, milk from a cow treated with genetically engineered bovine growth hormone would be a product 'produced with' genetically engineered material.

Specifically, my bill requires that food that contains or was produced with genetically engineered material be labeled at each stage of the food production process-- from seed company to farmer to manufacturer to retailer. The labeling requirement in my bill, however, does not to apply to drugs or to food sold in restaurants, bakeries, and other similar establishments.

Genetically engineered material is defined under the bill as material that 'has been altered at the molecular or cellular level by means that are not possible under natural conditions or processes.' Food developed through traditional processes such as crossbreeding is not considered to be genetically engineered, and the legislation's labeling requirement would not apply to foods produced in that way.

Under the bill, persons need not label food if they obtain a written guaranty from the party from whom they received the food that the food does not contain and was not produced with genetically engineered material. Persons who obtain a valid guaranty are not subject to penalties under the bill if they are later found to have failed to label food that contains genetically engineered material.

For example, a farmer who plants genetically engineered corn must label that corn. Each person who then buys and then sells that corn, or food derived from it, will also be required to label it as genetically engineered.

Conversely, farmers who obtain a guaranty that the corn they are planting is not genetically engineered may issue a guaranty to purchasers that their corn is not genetically engineered. The purchaser then would not have to label that corn or product made with that corn.

If the corn or food is later found to have contained or been produced with genetically engineered material but was not labeled accordingly, the purchaser would not be subject to penalties under the bill.

This guaranty system is used today to enforce provisions of existing law concerning the distribution of adulterated or mislabeled foods. The system is much less expensive than a system which would require food to be tested at every phase of the food production process.

Failure to label food that contains or was produced with genetically engineered material carries a civil penalty of up to $1,000 amount for each violation.

Importantly, the bill provides that if a party fraudulently warrants that a product is not genetically engineered, no party further down the chain of custody may be held liable for mislabeling. This provision is particularly meant to protect small farmers from the possibility that their suppliers would by contract provide that any liability for mislabeling be borne by the farmer regardless of the suppliers' own actions.

The bill also provides another protection for farmers. Under the bill, a farmer who plants a non-genetically engineered crop, but whose crop came to contain genetically engineered material from natural causes such as wind carrying pollen from a genetically engineered plant is not subject to penalties under the bill. This is the case so long as the farmer did not intend or did not negligently permit this to occur.

And, finally, the bill directs the Secretary of HHS to make grants to study the possible health and environmental risks associated with genetically engineered foods. The bill authorizes $5 million for this purpose.

In closing, Mr. President, during the recent negotiations on the Biosafety Protocol, it was the United States' negotiating position that international shipments of seeds, grains and plants that may contain genetically engineered material be labeled accordingly.

If the United States took the position that it is appropriate to provide this information to its trading partners, shouldn't we make similar information available to American consumers?

I am hopeful that my House and Senate colleagues can act quickly to ensure the passage of my legislation to give American families the right-to-know whether their food contains or was produced with genetically engineered material.

I ask that the text of my legislation be printed in the **Record.**

The text of the legislation follows:

S. 2080
Be it enacted by the Senate and House of Representatives of the United States of America in Congress assembled,

SECTION 1. SHORT TITLE.
This Act may be cited as the 'Genetically Engineered Food Right-to-Know Act'.

SEC. 2. FINDINGS.
Congress finds the following:

(1) In 1999, 98,600,000 acres in the United States were planted with genetically engineered crops, and more than 1/3 of the soybean crop, and 1/4 of the corn crop, in the United States was genetically engineered.

(2) The process of genetically engineering foods results in the material change of such foods.

(3) The health and environmental effects of genetically engineered foods are not yet known.

(4) Individuals in the United States have the right to know whether food contains or has been produced with genetically engineered material.

(5) Federal law gives individuals in the United States the right to know whether food contains artificial colors and flavors, chemical preservatives, and artificial sweeteners by requiring the labeling of such food.

(6) Requirements that genetically engineered food be labeled as genetically engineered would increase consumer knowledge about, and consumer control over consumption of, genetically engineered food.

(7) Genetically engineered material can be detected in food at levels as low as 0.1 percent by reasonably available technology.

SEC. 3. LABELING REGARDING GENETICALLY ENGINEERED MATERIAL; AMENDMENTS TO FEDERAL FOOD, DRUG, AND COSMETIC ACT.
(a) **In General**: Section 403 of the Federal Food, Drug, and Cosmetic Act (21 U.S.C. 343) is amended by adding at the end the following paragraph:
'(t)(1) If it contains a genetically engineered material, or was produced with a genetically engineered material, unless it bears a label (or labeling, in the case of a raw agricultural commodity) that provides notices in accordance with each of the following requirements:

'(A) The label or labeling bears the following notice: 'GENETICALLY ENGINEERED'.

'(B) The label or labeling bears the following notice: 'THIS PRODUCT CONTAINS A GENETICALLY ENGINEERED MATERIAL, OR WAS PRODUCED WITH A GENETICALLY ENGINEERED MATERIAL'.

'(C) The notice required in clause (A) immediately precedes the notice required in clause (B) and the type for the notice required in clause (A) is not less than twice the size of the type for the notice required in clause (B).

'(D) The notice required in clause (B) is the same size as would be required if the notice provided nutrition information that is required in paragraph (q)(1).

'(E) The notices required in clauses (A) and (B) are clearly legible and conspicuous.
'(2) This paragraph does not apply to food that--

'(A) is served in restaurants or other similar eating establishments, such as cafeterias and carryouts;

'(B) is a medical food as defined in section 5(b) of the Orphan Drug Act; or

'(C) was grown on a tree that was planted before the date of enactment of the Genetically Engineered Food Right-to-Know Act, in a case in which the producer of the food does not know if the food contains a genetically engineered material, or was produced with a genetically engineered material.
'(3) In this paragraph:

'(A) The term 'genetically engineered material' means material derived from any part of a genetically engineered organism, without regard to whether the altered molecular or cellular characteristics of the organism are detectable in the material.

'(B) The term 'genetically engineered organism' means--

'(i) an organism that has been altered at the molecular or cellular level by means that are not possible under natural conditions or processes (including recombinant DNA and RNA techniques, cell fusion, microencapsulation, macroencapsulation, gene deletion and doubling, introduction of a foreign gene, and a process that changes the positions of genes), other than a means consisting exclusively of breeding, conjugation, fermentation, hybridization, in vitro fertilization, or tissue culture; and

'(ii) an organism made through sexual or asexual reproduction, or both, involving an organism described in subclause (i), if possessing any of the altered molecular or cellular characteristics of the organism so described.

'(C) The term 'produced with a genetically engineered material', used with respect to a food, means a food if--

'(i) the organism from which the food is derived has been injected or otherwise treated with a genetically engineered material (except that the use of manure as a fertilizer for raw agricultural commodities may not be construed to be production with a genetically engineered material);

'(ii) the animal from which the food is derived has been fed genetically engineered material; or

'(iii) the food contains an ingredient that is a food to which subclause (i) or (ii) applies.'.
(b) **Guaranty**:

(1) **In general**: Section 303(d) of the Federal Food, Drug, and Cosmetic Act (21 U.S.C. 333(d)) is amended--

(A) by striking '(d)' and inserting '(d)(1)'; and

(B) by adding at the end the following paragraph:
'(2)(A) No person shall be subject to the penalties of subsection (a)(1) or (h) for a violation of section 301(a), 301(b), or 301(c) involving food that is misbranded within the meaning of section 403(t) if such person (referred to in this paragraph as the 'recipient') establishes a guaranty or undertaking that--

'(i) is signed by, and contains the name and address of, a person residing in the United States from whom the recipient received in good faith the food (including the receipt of seeds to grow raw agricultural commodities); and

'(ii) contains a statement to the effect that the food does not contain a genetically engineered material or was not produced with a genetically engineered material.
'(B) In the case of a recipient who, with respect to a food, establishes a guaranty or undertaking in accordance with subparagraph (A), the exclusion under such subparagraph from being subject to penalties applies to the recipient without regard to the manner in which the recipient uses the food, including whether the recipient is--

'(i) processing the food;

'(ii) using the food as an ingredient in a food product;

'(iii) repacking the food; or

'(iv) growing, raising, or otherwise producing the food.
'(C) No person may avoid responsibility or liability for a violation of section 301(a), 301(b), or 301(c) involving food that is misbranded within the meaning of section 403(t) by entering into a contract or other agreement that specifies that another person shall bear such responsibility or liability, except that a recipient may require a guaranty or undertaking as described in this subsection.
'(D) In this paragraph, the terms 'genetically engineered material' and 'produced with a genetically engineered material' have the meanings given the terms in section 403(t).'.

(2) **False guaranty**: Section 301(h) of the Federal Food, Drug, and Cosmetic Act (21 U.S.C. 331(h)) is amended by inserting 'or 303(d)(2)' before ', which guaranty or undertaking is false' the first place it appears.
(c) **Unintended Contamination**: Section 303(d) of the Federal Food, Drug, and Cosmetic Act, as amended by subsection (b)(1), is further amended by adding at the end the following paragraph:
'(3)(A) No person shall be subject to the penalties of subsection (a)(1) or (h) for a violation of section 301(a), 301(b), or 301(c) involving food that is misbranded within the meaning of section 403(t) if--

'(i) such person is an agricultural producer and the violation occurs because food that is grown, raised, or otherwise produced by such producer, which food does not contain a genetically engineered material and was not produced with a genetically engineered material, is contaminated with a food that contains a genetically engineered material or was produced with a genetically engineered material (including contamination by mingling the 2 foods); and

'(ii) such contamination is not intended by the agricultural producer.
'(B) Subparagraph (A) does not apply to an agricultural producer to the extent that the contamination occurs as a result of the negligence of the producer.'.
(d) **Civil Penalties**: Section 303 of the Federal Food, Drug, and Cosmetic Act (21 U.S.C. 333) is amended by adding at the end the following subsection:
'(h)(1) With respect to a violation of section 301(a), 301(b), or 301(c) involving food that is misbranded within the meaning of section 403(t), any person engaging in such a violation shall be liable to the United States for a civil penalty in an amount not to exceed $1,000 for each such violation.
'(2) Paragraphs (3) through (5) of subsection (g) apply with respect to a civil penalty assessed under paragraph (1) to the same extent and in the same manner as such paragraphs (3) through (5) apply with respect to a civil penalty assessed under paragraph (1) or (2) of subsection (g).'.

SEC. 4. GRANTS FOR RESEARCH ON GENETICALLY ENGINEERED FOOD.
Chapter IX of the Federal Food, Drug, and Cosmetic Act (21 U.S.C. 391 et seq.) is amended by adding at the end the following:

'SEC. 908. GRANTS FOR RESEARCH ON GENETICALLY ENGINEERED FOOD.

'(a) **In General:** The Secretary may make grants to appropriate individuals, organizations, and institutions to conduct research into the public health and environmental risks associated with genetically engineered materials, food that contains a genetically engineered material, and food that is produced with a genetically engineered material, including risks related to--

'(1) increased allergenicity;

'(2) increased toxicity;

'(3) cross-pollination between genetically engineered materials and materials that are not genetically engineered materials; and

'(4) interference with the soil ecosystem and other impacts on the ecosystem.
'(b) **Authorization of Appropriations:**

'(1) **In general**: There is authorized to be appropriated $5,000,000 for fiscal year 2001 to carry out the objectives of this section.

'(2) **Availability**: Any sums appropriated under the authorization contained in this subsection shall remain available, without fiscal year limitation, until expended.
'(c) **Definitions:** The terms 'genetically engineered material' and 'produced with a genetically engineered material' have the meanings given the terms in section 403(t)(3) of the Federal Food, Drug, and Cosmetic Act.'.

SEC. 5. CONFORMING AMENDMENTS.
(a) Section 1(n) of Public Law 90-201 is amended--

(1) in paragraph (11), by striking 'or' at the end;

(2) in paragraph (12), by striking the period at the end and inserting '; or'; and

(3) by adding at the end the following:

'(13) if--

'(A) it contains a genetically engineered material, or was produced with a genetically engineered material; and

'(B)(i) it does not bear a label or labeling, as appropriate, that provides the notices required under the terms and conditions of section 403(t) of the Federal Food, Drug, and Cosmetic Act (21 U.S.C. 343(t)); or

'(ii) it is the subject of a false guaranty or undertaking,

subject to the terms and conditions of section 303(d) of that Act (21 U.S.C. 333(d)) and subject to the penalties described in section 303(h) of that Act (21 U.S.C. 333(h)) and remedies available under this Act.'.
(b) Section 4(h) of Public Law 85-172 is amended--

(1) in paragraph (11), by striking 'or' at the end;

(2) in paragraph (12), by striking the period at the end and inserting '; or'; and

(3) by adding at the end the following:

'(13) if--

'(A) it contains a genetically engineered material, or was produced with a genetically engineered material; and

'(B)(i) it does not bear a label or labeling, as appropriate, that provides the notices required under the terms and conditions of section 403(t) of the Federal Food, Drug, and Cosmetic Act (21 U.S.C. 343(t)); or

'(ii) it is the subject of a false guaranty or undertaking,

subject to the terms and conditions of section 303(d) of that Act (21 U.S.C. 333(d)) and subject to the penalties described in section 303(h) of that Act (21 U.S.C. 333(h)) and remedies available under this Act.'.

SEC. 6. EFFECTIVE DATE.
This Act and the amendments made by this Act take effect 180 days after the date of enactment of this Act.

Congressional Record May 26, 1999 (Edited)

Mr. KUCINICH: Mr. Chairman, a few years ago I visited an elementary school in Cleveland at the start of the school year. The children celebrating the beginning of their school year had released hundreds and hundreds of butterflies into the air.

Now, a butterfly is a powerful symbol in our society. It is a symbol of transformation, transformation from a caterpillar into this beautiful winged being. Butterflies excite the imagination, they enthrall us with their possibilities. Yet, the butterfly may become the next casualty of our brave new world.

We are all familiar with the genetically altered crops where pesticides are engineered right into the crop. A recent study indicates that pollen from such crops may have the potential to kill off butterflies, including the majestic and beautiful Monarch butterfly.

Mr. Chairman, my intention with this amendment is to provide the Agricultural Research Service with $100,000 to study the effects of pollen from genetically modified crops on harmless insects, and to study the effect on other species, including animals and humans, that may come in contact with the pollen.

Corn that has been genetically engineered with the pesticide Bt has been approved and was introduced to farmers' fields in 1996. It now accounts for one-fourth of the Nation's corn crop. Bt is toxic to European and Southwestern corn borers, caterpillars that mine into corn stalks and destroy developing ears of corn.

According to a recent study conducted at Cornell University, it is also deadly to Monarch butterflies. The Cornell study found that after feeding a group of larvae, milkweed leaves dusted with Bt pollen, almost half died. The larvae that did survive were small and lethargic.

The implications of this are very clear. Pollen from Bt-exuding corn spreads to milkweed plants, which grow around the edges of cornfields. Monarch larvae feed exclusively on milkweed. Every year, Monarchs migrate from Mexico and southern States, and many of them grow from caterpillars into beautiful black, orange, and white butterflies in the United States corn belt during the time the corn pollination occurs.

I am sure that millions of Americans have had the experience of taking their children in hand and going into a pasture and watching for beautiful butterflies to come by and visiting an arboretum, a zoo, a park and watching the butterflies.

Well, now, if we read the Washington Post, it says that pollen from plants can blow onto nearby milkweed plants, the exclusive food upon which the Monarch larvae feed, and get eaten by the tiger-striped caterpillars.

At laboratory studies at Cornell, the engineered pollen killed nearly half of those young before they transformed into the brilliant orange, black, and white butterflies so well-known throughout North America. Several scientists expressed concern that if the new study results are correct, then monarchs, which already face ecological pressures, but so far have managed to hold their own, may soon find themselves on the Endangered Species list. Other butterflies may soon be at risk.

From the Friends of the Earth we hear, "The failure of Congress and the administration to ensure more careful control over genetically modified organisms has unleashed a frightening experiment on the people and environment of the United States. It is time to look more closely at the flawed review process of the three Federal agencies that regulate genetically modified products: EPA, FDA, and USDA.

"The implications of the Cornell University study go far beyond Monarch butterflies and point to the need for a revamping of our regulatory framework on biotechnology."

Monarchs have already lost much of their habitat when tall-grass prairies were converted to farmland. We now need to protect them and other species that are harmless to farmers' crops, that may be adversely affected by Bt pollen.

It is shocking that more extensive studies like the one performed at Cornell were not done before the crop was approved. It also makes one wonder what effects other genetically altered crops may have on other species, such as birds, bees, and even humans, and if adequate risk assessments are being done on bioengineered products before they are approved and released into the environment.

My fellow colleagues, more research obviously needs to be done on these transgenic crops. I ask my colleagues to support my amendment to protect Monarch butterflies from the harmful effects of genetically modified crops.

Finally, Mr. Chairman, last year I had the opportunity to visit Pelee Island in Canada, which is a migration point for the Monarch butterflies. There is nothing more beautiful than to see hundreds of thousands of these beautiful creatures moving in a migratory pattern. It is an awesome sight. And yet, because of a lack of foresight on the part of our government, there is the possibility that these beautiful creatures may in fact be doomed. That is why this amendment is important.

NEXT CONGRESSWOMAN KAPTUR SPOKE

Ms. KAPTUR: Mr. Chairman, I rise in strong support of the amendment dealing with research by the Agricultural Research Service for the Monarch butterfly. Let me just say that the Committee on Agriculture, which the gentleman from New Mexico (Mr. Skeen) chairs and of which I am the ranking member, is the chief ecosystem committee of this Congress, and I believe, of this country.

There is an expression: "You can't fool Mother Nature." There are some fundamental questions being raised here by the gentleman from Ohio (Mr. Kucinich) that are very important to the future of botanical life and biological life in our country. Because we have never before had these genetically engineered crops, we really do not know their long-term impacts.

I know recent articles in Scientific American and many newspapers indicate that as a result of butterflies, which are essential to pollinating crops so we can produce fruit and corn, and representing the eastern part of the eastern corn belt, we know something about corn and soybeans, and these butterflies are essential to our future. After being impacted by this pollen, 40 percent of them died. 40 percent. This is a profound result. So I think the gentleman from Ohio (Mr. Kucinich) brings to us a very important and current finding that is well deserving of research.

I also would say to the gentleman, I thank him for doing this, because I know he represents the inner part of Cleveland, Ohio; and one of my greatest concerns as another American is that we have the first generation of Americans now that have no connection to the land. We have literally raised the first generation of people in the Nation's history who do not spend the majority of their time raising their food or with any connection to production at all, so they are divorced from the experiences that he is talking about.

I would just say, for someone from Cleveland, Ohio, a major city in this country, to bring this amendment to the floor, to me, in some ways is a modern-day miracle. So I want to thank the gentleman, and I look forward to supporting him.

Mr. KUCINICH: Mr. Chairman, will the gentlewoman yield?

Ms. KAPTUR: I yield to the gentleman from Ohio.

Mr. KUCINICH: Mr. Chairman, I appreciate the gentlewoman's response. And it is an honor to serve with the gentlewoman in this Congress, serving the people of Ohio.

She raised an interesting point, and that is, what effect do these genetically engineered products have on our natural environment? I mean, sometime in the 20th century there was kind of a disconnection between humanity and the natural environment; and we will spend, I suppose, a good part of the next century trying to reconnect.

The disassociation from the land which the gentlewoman speaks about is a profound disconnection from nature. I think that is why schoolchildren, for example, find it so fascinating to study butterflies. Because in some ways, that primal human sympathy which Wordsworth talked about in his poetry flutters in the heart when we see something so beautiful. And I think that as the schoolchildren, who spend time with their parents and their grandparents going to parks and zoos and arboretums, have the knowledge that this very beautiful butterfly could be impacted by this bioengineering, I think that we are going to see a response nationally. And it would be healthy because this country needs to look for opportunities to reconnect with our natural state.

So I thank the gentlewoman. I would hope that the esteemed chairman, the gentleman from New Mexico (Mr. Skeen) would be able to respond.

Mr. SKEEN: Mr. Chairman, will the gentlewoman yield?

Ms. KAPTUR: I yield to the gentleman from New Mexico.

Mr. SKEEN: Mr. Chairman, I will tell the gentleman I am all aflutter. I would like to say that I understand the concern of the gentleman, and I will continue to work with him to address this situation, and I think he has got a good program.

Mr. KUCINICH: Mr. Chairman, if the gentlewoman would continue to yield, I would be more than happy to work with the chair. I need the help of the gentlewoman from Ohio (Ms. Kaptur) and I need the help of the Chair. We can work together to address this issue, bring it to the committee.

With that kind of assurance, I say to the gentleman from New Mexico (Mr. Skeen), I will withdraw the amendment, but look forward to working with both of my colleagues to find the appropriate venue within the committee so that we can start to get these agencies to be aware of this major concern of public policy.

I thank the gentleman again for his work on this matter and for his work on the agricultural bill. And again, my gratitude to the gentlewoman from Ohio (Ms. Kaptur). It is an honor to be with her in this House.

Ms. KAPTUR: Mr. Chairman, I say to the gentleman from Cleveland, Ohio (Mr. Kucinich) that I thank him very much for bringing this to the Nation's attention. He is a leader on this issue, and I look forward to working with our chairman to find an answer to this as we move toward the conference.

---

http://www.thecampaign.org/kucinick52699.htm

106TH CONGRESS
2D SESSION

H. R. 3883

To amend the Federal Food, Drug, and Cosmetic Act with respect to the safety of genetically engineered foods.

IN THE HOUSE OF REPRESENTATIVES

MARCH 9, 2000

Mr. KUCINICH (for himself, Mr. METCALF, Mr. HINCHEY, Mr. CONYERS, Mr. SANDERS, Ms. WOOLSEY, and Ms. LEE) introduced the following bill; which was referred to the Committee on Commerce

---

A BILL

To amend the Federal Food, Drug, and Cosmetic Act with respect to the safety of genetically engineered foods.

Be it enacted by the Senate and House of Representatives of the United States of America in Congress assembled,

SECTION 1. SHORT TITLE.

This Act may be cited as the "Genetically Engineered Food Safety Act".

SEC. 2. FINDINGS.

The Congress finds as follows:

(1) Genetic engineering is an artificial gene transfer process wholly different from traditional breeding.

(2) Genetic engineering can be used to produce new versions of virtually all plant and animal foods. Thus, within a short time, the food supply could consist almost entirely of genetically engineered products.

(3) This conversion from a food supply based on traditionally bred organisms to one based on organisms produced through genetic engineering could be one the most important changes in our food supply in this century.

(4) Genetically engineered foods present new issues of safety that have not been adequately studied.

(5) The Congress has previously required that food additives be analyzed for their safety prior to their placement on the market.

(6) Adding new genes into a food should be considered adding a food additive, thus requiring an analysis of safety factors.

(7) Federal agencies have failed to uphold congressional intent of the Food Additives Amendment of 1958 by allowing genetically engineered foods to be marketed, sold and otherwise used without requiring pre-market safety testing addressing their unique characteristics.

(8) The food additive process gives the Food and Drug Administration discretion in applying the safety factors that are generally recognized as appropriate to evaluate the safety of food and food ingredients.

SEC. 3. FEDERAL DETERMINATION OF SAFETY OF GENETICALLY ENGINEERED FOOD; REGULATION AS FOOD ADDITIVE.

(a) INCLUSION IN DEFINITION OF FOOD ADDITIVE.—Section 201 of the Federal Food, Drug, and Cosmetic Act (21 U.S.C. 321) is amended—

(1) in paragraph (s), by adding after and below subparagraph (6) the following sentence:

"Such term includes the different genetic constructs, proteins of such constructs, vectors, promoters, marker systems, and other appropriate terms that are used or created as a result of the creation of a genetically engineered food (as defined in paragraph (kk)), other than a genetic construct, protein, vector, promoter, or marker system or other appropriate term for which an application under section 505 or 512 has been filed. For purposes of this Act, the term 'genetic food additive' means a genetic construct, protein, vector, promoter, or marker system or other appropriate term that is so included."; and

(2) by adding at the end the following:

"(kk)(1) The term 'genetically engineered food' means food that contains or was produced with a genetically engineered material.

"(2) The term 'genetically engineered material' means material derived from any part of a genetically engineered organism, without regard to whether the altered molecular or cellular characteristics of the organism are detectable in the material.

"(3) The term 'genetically engineered organism' means—

"(A) an organism that has been altered at the molecular or cellular level by means that are not possible under natural conditions or processes (including but not limited to recombinant DNA and RNA techniques, cell fusion, microencapsulation, macroencapsulation, gene deletion and doubling, introducing a foreign gene, and changing the positions of genes), other than a means consisting exclusively of breeding, conjugation, fermentation, hybridization, in vitro fertilization, or tissue culture, and

"(B) an organism made through sexual or asexual reproduction (or both) involving an organism described in clause (A), if possessing any of the altered molecular or cellular characteristics of the organism so described.

"(4) For purposes of subparagraph (1), a food shall be considered to have been produced with a genetically engineered material if the organism from which the food is derived has been injected or otherwise treated with a genetically engineered material (except that the use of manure as a fertilizer for raw agricultural commodities may not be construed to mean that such commodities are produced with a genetically engineered material).".

(b) PETITION TO ESTABLISH SAFETY.—

(1) DATA IN PETITION.—Section 409(b)(2)(E) of the Federal Food, Drug, and Cosmetic Act (21 U.S.C. 348(b)(2)(E)) is amended by adding at the end the following sentence: "In the case of a genetic food additive, such reports shall include all data that was collected or developed pursuant to the investigations, including data that does not support the claim of safety for use.".

(2) NOTICES; PUBLIC AVAILABILITY OF INFORMATION.—Section 409(b)(5) of the Federal Food, Drug, and Cosmetic Act (21 U.S.C. 348(b)(5)) is amended—

(A) by striking "(5)" and inserting "(5)(A)"; and

(B) by adding at the end the following subparagraphs:

"(B) In the case of a genetic food additive, the Secretary, promptly after providing the notice under subparagraph (A), shall make available to the public all reports and data described in paragraph (2)(E) that are contained in the petition involved, and all other information in the petition to the extent that the information is relevant to a determination of the safety for use of the additive. Such notice shall state whether any information in the petition is not being made available to the public because the Secretary has made a determination that the information does not relate to the safety for use of the additive. Any person may petition the Secretary for a reconsideration of such a determination, and if the Secretary finds in favor of such person, the period for public comment under subsection (c)(2)(B) shall be extended accordingly.

"(C) In the case of genetic food additives:

"(i) The Secretary shall maintain and make available to the public through telecommunications a list of petitions that are pending under this subsection and a list of petitions for which regulations under subsection (c)(1)(A) have been established. Such list shall include information on the additives involved, including the source of the additives, and including any information received by the Secretary pursuant to clause (ii).

"(ii) If a regulation is in effect under subsection (c)(1)(A) for a genetic food additive, any person who manufactures such additive for commercial use shall submit to the Secretary a notification of any knowledge of data that relate to the adverse health effects of the additive, when knowledge is acquired by the person after the date on which the regulation took effect. If the manufacturer is in possession of the data, the notification shall include the data. The Secretary shall by regulation establish the scope of the responsibilities of manufacturers under this clause, including such limits on the responsibilities as the Secretary determines to be appropriate.".

(3) EFFECTIVE DATE OF REGULATION REGARDING SAFE USE; OPPORTUNITY FOR PUBLIC COMMENT.—Section 409(c)(2) of the Federal Food, Drug, and Cosmetic Act (21 U.S.C. 348(c)(2)) is amended—

(A) by striking "(2)" and inserting "(2)(A)"; and

(B) by adding at the end the following subparagraph:

"(B) In the case of a genetic food additive, an order under paragraph (1)(A) may not be issued before the expiration of the 30-day period beginning on the date on which the Secretary has under subsection (b)(5) made information available to the public pursuant to a notification under such subsection regarding the petition involved. During such period (or such longer period as the Secretary may designate), the Secretary shall provide interested persons an opportunity to submit to the Secretary comments on the petition. In publishing such notice, the Secretary shall inform the public of such opportunity.".

(3) CONSIDERATION OF CERTAIN FACTORS.—Section 409(c) of the Federal Food, Drug, and Cosmetic Act (21 U.S.C. 348(c)) is amended by adding at the end the following paragraph:

"(6) In the case of a genetic food additive, the factors considered by the Secretary regarding safety for use shall include (but not be limited to) the results of the following analyses:

"(A) Allergenicity effects resulting from the added proteins, including proteins not found in the food supply.

"(B) Pleiotropic effects. The Secretary shall require tests to determine the potential for such effects (using molecular characterization, biochemical characterization, mRNA profiling, or other techniques, or as appropriate, combinations of such techniques).

"(C) Appearance of new toxins or increased levels of existing toxins.

"(D) Changes in the functional characteristics of food.

"(E) Changes in the levels of important nutrients.".

(4) CERTAIN TESTS.—Section 409(c) of the Federal Food, Drug, and Cosmetic Act, as amended by paragraph (3), is amended by adding at the end the following paragraph:

"(7) In the case of genetic food additives:

"(A) If a genetic food additive is a protein from a commonly or severely allergenic food, the Secretary may not establish a regulation under paragraph (1)(A) if the petition under subsection (b)(1) fails to include full reports of investigations that used serum or skin tests (or other advanced techniques) on a sensitive population to determine whether such additive is commonly or severely allergenic.

"(B)(i) If a genetic food additive is a protein that has not undergone the investigations described in subparagraph (A), the Secretary may not establish a regulation under paragraph (1)(A) if the petition under subsection (b)(1) fails to include full reports of investigations that used the best available biochemical and physiological protocols to evaluate whether it is likely that the protein involved is an allergen.

"(ii) For purposes of clause (i), the Secretary shall by regulation determine the best available biochemical and physiological protocols. In carrying out rulemaking under the preceding sentence, the Secretary shall consult with the Director of the National Institutes of Health.".

(5) PROHIBITED ADDITIVES.—Section 409(c) of the Federal Food, Drug, and Cosmetic Act, as amended by paragraph (4), is amended by adding at the end the following paragraph:

"(8) In the case of a genetic food additive, the Secretary may not establish a regulation under paragraph (1)(A) if—

"(A) the additive is a protein and a report of an investigation finds that the additive is likely to be commonly or severely allergenic;

"(B) the additive is a protein and a report of an investigation that uses a protocol described in paragraph (7)(B) fails to find with reasonable certainty that the additive is unlikely to be an allergen; or

"(C) effective June 1, 2004, a selective marker is used with respect to the additive, the selective marker will remain in the food involved when the food is marketed, and the selective marker inhibits the function of one or more antibiotics.".

(6) ADDITIONAL PROVISIONS.—Section 409(c) of the Federal Food, Drug, and Cosmetic Act, as amended by paragraph (5), is amended by adding at the end the following paragraph:

"(9)(A) In determining the safety for use of genetic food additives, the Secretary may (directly or through contract) conduct investigations of such additives for purposes of supplementing the information provided to the Secretary pursuant to petitions under subsection (b)(1).

"B) To provide the Congress with a periodic independent, external review of the Secretary's formulation of the approval process under paragraph (1)(A) that relates to genetic food additives, the Secretary shall enter into an agreement with the Institute of Medicine. Such agreement shall provide that, if the Institute of Medicine has any concerns regarding the approval process, the Institute of Medicine will submit to the Congress a report describing such concerns.

"(C) In the case of genetic food additives, petitions under subsection (b)(1) may not be categorically excluded for purposes of the National Environmental Policy Act.".

(c) REGULATION ISSUED ON SECRETARY'S INITIATIVE.—Section 409(d) of the Federal Food, Drug, and Cosmetic Act (21 U.S.C. 348(d)) is amended—

(1) by striking "(d) The Secretary" and inserting "(d)(1) Subject to paragraph (2), the Secretary"; and

(2) by adding at the end the following paragraph:

"(2) The provisions of subsections (b) and (c) that expressly reference genetic food additives apply with respect to a regulation proposed by the Secretary under paragraph (1) to the same extent and in the same manner as such provisions apply with respect to a petition filed under subsection (b)(1).".

(d) CIVIL PENALTIES.—Section 303 of the Federal Food, Drug, and Cosmetic Act (21 U.S.C. 333) is amended by adding at the end the following subsection:

"(h)(1) With respect to a violation of section 301(a), 301(b), or 301(c) involving the adulteration of food by reason of failure to comply with the provisions of section 409 that relate to genetic food additives, any person engaging in such a violation shall be liable to the United States for a civil penalty in an amount not to exceed $100,000 for each such violation.

"(2) Paragraphs (3) through (5) of subsection (g) apply with respect to a civil penalty under paragraph (1) of this subsection to the same extent and in the same manner as such paragraphs (3) through (5) apply with respect to a civil penalty under paragraph (1) or (2) of subsection (g).".

(e) RULE OF CONSTRUCTION.—With respect to section 409 of the Federal Food, Drug, and Cosmetic Act as amended by this section, compliance with the provisions of such section 409 that relate to genetic food additives does not constitute an affirmative defense in any cause of action under Federal or State law for personal injury resulting in whole or in part from a genetic food additive.

SEC. 4. USER FEES REGARDING DETERMINATION OF SAFETY OF GENETIC FOOD ADDITIVES.

Chapter IV of the Federal Food, Drug, and Cosmetic Act (21 U.S.C. 341 et seq.) is amended by inserting after section 409 the following section:

"USER FEES REGARDING SAFETY OF GENETIC FOOD ADDITIVES

"SEC. 409A. (a) IN GENERAL.—In the case of genetic food additives, the Secretary shall in accordance with this section assess and collect a fee on each petition that is filed under section 409(b)(1). The fee shall be collected from the person who submits the petition, is due upon submission of the petition, and shall be assessed in an amount determined under subsection (c). This section applies as of the first fiscal year that begins after the date of promulgation of the final rule required in section 5 of the Genetically Engineered Food Safety Act (referred to in this section as the 'first applicable fiscal year').

"(b) PURPOSE OF FEES.—

"(1) IN GENERAL.—The purposes of fees under subsection (a) are as follows:

"(A) To defray increases in the costs of the resources allocated for carrying out section 409 for the first applicable fiscal year over the costs of carrying out such section for the preceding fiscal year, other than increases that are not attributable to the responsibilities of the Secretary with respect to genetic food additives.

"(B) To provide for a program of basic and applied research on the safety of genetic food additives (to be carried out by the Commissioner of Food and Drugs). The program shall address fundamental questions and problems that arise repeatedly during the process of reviewing petitions under section 409(b)(1) with respect to genetic food additives, and shall not directly support the development of new genetically engineered foods.

"(2) ALLOCATIONS BY SECRETARY.—Of the total fee revenues collected under subsection (a) for a fiscal year, the Secretary shall reserve and expend—

"(A) 95 percent for the purpose described in paragraph (1)(A) and

"(B) 5 percent for the purpose described in paragraph (1)(B).

"(3) CERTAIN PROVISIONS REGARDING INCREASED ADMINISTRATIVE COSTS.—With respect to fees under subsection (a):

"(A) Increases referred to in paragraph (1)(A) include the costs of the Secretary in providing for investigations under section 409(c)(9)(A).

"(B) Increases referred to in paragraph (1)(A) include increases in costs for an additional number of full-time equivalent positions in the Department of Health and Human Services to be engaged in carrying out section 409 with respect to genetic food additives.

"(c) TOTAL FEE REVENUES; INDIVIDUAL FEE AMOUNTS.—The total fee revenues collected under subsection (a) for a fiscal year shall be the amounts appropriated under subsection (f)(2) for such fiscal year. Individual fees shall be assessed by the Secretary on the basis of an estimate by the Secretary of the amount necessary to ensure that the sum of the fees collected for such fiscal year equals the amount so appropriated.

"(d) FEE WAIVER OR REDUCTION.—The Secretary shall grant a waiver from or a reduction of a fee assessed under subsection (a) if the Secretary finds that the fee to be paid will exceed the anticipated present and future costs incurred by the Secretary in carrying out the purposes described in subsection (b) (which finding may be made by the Secretary using standard costs).

"(e) ASSESSMENT OF FEES.—

"(1) LIMITATION.—Fees may not be assessed under subsection (a) for a fiscal year beginning after the first applicable fiscal year unless the amount appropriated for salaries and expenses of the Food and Drug Administration for such fiscal year is equal to or greater than the amount appropriated for salaries and expenses of the Food and Drug Administration for the first applicable fiscal year multiplied by the adjustment factor applicable to the fiscal year involved, except that in making determinations under this paragraph for the fiscal years involved there shall be excluded—

"(A) the amounts appropriated under subsection (f)(2) for the fiscal years involved; and

"(B) the amounts appropriated under section 736(g) for such fiscal years.

"(2) AUTHORITY.—If under paragraph (1) the Secretary does not have authority to assess fees under subsection (a) during a portion of a fiscal year, but does at a later date in such fiscal year have such authority, the Secretary, notwithstanding the due date under such subsection for fees, may assess and collect such fees at any time in such fiscal year, without any modification in the rate of the fees.

"(f) CREDITING AND AVAILABILITY OF FEES.—

"(1) IN GENERAL.—Fees collected for a fiscal year pursuant to subsection (a) shall be credited to the appropriation account for salaries and expenses of the Food and Drug Administration and shall be available in accordance with appropriation Acts until expended without fiscal year limitation. Such sums as may be necessary may be transferred from the Food and Drug Administration salaries and expenses appropriation account without fiscal year limitation to such appropriation account for salaries and expenses with such fiscal year limitation. The sums transferred shall be available solely for the purposes described in paragraph (1) of subsection (b), and the sums are subject to allocations under paragraph (2) of such subsection.

"(2) AUTHORIZATION OF APPROPRIATIONS.—

"(A) FIRST FISCAL YEAR.—For the first applicable fiscal year—

"(i) there is authorized to be appropriated for fees under subsection (a) an amount equal to the amount of increase determined under subsection (b)(1) by the Secretary (which amount shall be published in the Federal Register); and

"(ii) in addition, there is authorized to be appropriated for fees under subsection (a) an amount determined by the Secretary to be necessary to carry out the purpose described in subsection (b)(2) (which amount shall be so published).

"(B) SUBSEQUENT FISCAL YEARS.—For each of the four fiscal years following the first applicable fiscal year—

"(i) there is authorized to be appropriated for fees under subsection (a) an amount equal to the amount that applied under subparagraph (A)(i) for the first applicable fiscal year, except that such amount shall be adjusted under paragraph (3)(A) for the fiscal year involved; and

"(ii) in addition, there is authorized to be appropriated for fees under subsection (a) an amount equal to the amount that applied under subparagraph (A)(ii) for the first applicable fiscal year, except that such amount shall be adjusted under paragraph (3)(B) for the fiscal year involved.

"(3) ADJUSTMENTS.—

"(A) AGENCY COST OF RESOURCES.—For each fiscal year other than the first applicable fiscal year, the amount that applied under paragraph (2)(A)(i) for the first applicable fiscal year shall be multiplied by the adjustment factor (as defined in subsection (i)).

"(B) RESEARCH PROGRAM.—For each fiscal year other than the first applicable fiscal year, the amount that applied under paragraph (2)(A)(ii) for the first applicable fiscal year shall be adjusted by the Secretary (and as adjusted shall be published in the Federal Register) to reflect the greater of—

"(i) the total percentage change that occurred during the preceding fiscal year in the Consumer Price Index for all urban consumers (all items; U.S. city average); or

"(ii) the total percentage change for such fiscal year in basic pay under the General Schedule in accordance with section 5332 of title 5, United States Code, as adjusted by any locality-based comparability payment pursuant to section 5304 of such title for Federal employees stationed in the District of Columbia.

"(4) OFFSET.—Any amount of fees collected for a fiscal year under subsection (a) that exceeds the amount of fees specified in appropriation Acts for such fiscal year shall be credited to the appropriation account of the Food and Drug Administration as provided in paragraph (1), and shall be subtracted from the amount of fees that would otherwise be authorized to be collected under this section pursuant to appropriation Acts for a subsequent fiscal year.

"(g) COLLECTION OF UNPAID FEES.—In any case where the Secretary does not receive payment of a fee assessed under subsection (a) within 30 days after it is due, such fee shall be treated as a claim of the United States Government subject to subchapter II of chapter 37 of title 31, United States Code.

"(h) CONSTRUCTION.—This section may not be construed as requiring that the number of full-time equivalent positions in the Department of Health and Human Services, for officers, employers, and advisory committees not engaged in carrying out section 409 with respect to genetic food additives be reduced to offset the number of officers, employees, and advisory committees so engaged.

"(i) DEFINITION OF ADJUSTMENT FACTOR.—For purposes of this section, the term 'adjustment factor' applicable to a fiscal year is the lower of—

"(1) the Consumer Price Index for all urban consumers (all items; United States city average) for April of the preceding fiscal year divided by such Index for April of the first applicable fiscal year; or

"(2) the total of discretionary budget authority provided for programs in categories other than the defense category for the immediately preceding fiscal year (as reported in the Office of Management and Budget sequestration preview report, if available, required under section 254(c) of the Balanced Budget and Emergency Deficit Control Act of 1985) divided by such budget authority for the first applicable fiscal year (as reported in the Office of Management and Budget final sequestration report submitted for such year).

For purposes of this subsection, the terms 'budget authority' and 'category' have the meaning given such terms in the Balanced Budget and Emergency Deficit Control Act of 1985.".

SEC. 5. RULEMAKING; EFFECTIVE DATE; PREVIOUSLY UNREGULATED MARKETED ADDITIVES.

(a) RULEMAKING; EFFECTIVE DATE.—Not later than one year after the date of the enactment of this Act, the Secretary of Health and Human Services shall by regulation establish criteria for carrying out section 409 of the Federal Food, Drug, and Cosmetic Act in accordance with the amendments made by section 3, and criteria for carrying out section 409A of such Act (as added by section 4). Such amendments take effect upon the expiration of the 30-day period beginning on the date on which the Secretary promulgates the final rule under the preceding sentence, subject to subsection (b).

(b) PREVIOUSLY UNREGULATED MARKETED ADDITIVES.—

(1) IN GENERAL.—In the case of a genetic food additive (as defined pursuant to the amendments made by section (3)) that in the United States was in commercial use in food as of the day before the date on which the final rule under subsection (a) is promulgated, the amendments made by this Act apply to the additive upon the expiration of the two-year period beginning on the date on which the final rule is promulgated, subject to paragraph (2).

(2) USER FEES.—With respect to a genetic food additive described in paragraph (1), such paragraph does not waive the applicability of section 409A of the Federal Food, Drug, and Cosmetic Act to a petition under section 409(b)(1) of such Act that is filed before the expiration of the two-year period described in such paragraph.

http://www.thecampaign.org/HR3883.htm

Section II

Regulation Documents

The documents in this section include not only position papers presented by interest groups to governmental agencies on proposed rules, but other materials prepared and distributed to the public by government agencies themselves.

The first four are government documents: a speech delivered by then Secretary of Agriculture Dan Glickman to the National Press Club on July 13, 1999; an interview with then Commissioner of the Food and Drug Administration Jane Henney on the issues raised by bioengineered foods, published in the FDA's *Consumer Magazine* in early 2000; a fact sheet issued by the U.S. Department of State on March 21, 2000 on the importance of bioengineered food in solving hunger problems here and abroad; and a formal announcement by the Department of Health and Human Services through the Office of the White House that the FDA will strengthen its regulatory protocol on genetically modified foods.

The next piece is a detailed description of the process by which plant biotechnology is regulated, prepared by the American Crop Protection Association. It is followed by a sample of responses by interest groups to the FDA's proposed rules governing genetically engineered foods: from the Center for Food Safety, the Grocery Manufacturers of America, Inc., the American Farm Bureau, the Environmental Research Foundation, Greenpeace, and the Competitive Enterprise Institute.

Discussion

The government documents reproduced here, though not technically part of the regulatory process, demonstrate that administrative agencies are not dispassionate participants in the regulatory process but, in an important sense, advocates themselves. While paying lip service to the need to assure the public of the safety and efficacy of genetically altered foods and to the principle of informed consent, all betray a clear bias in favor of the ultimate benefits of the technology as here applied. The Department of Agriculture in particular is saddled with a dual — and potentially conflicting — mission. It must, of course, remain true to its solemn obligation to assure the safety of the nation's food supply. At the same time, an aspect of this mission is to assure the citizens of the country of the sufficiency of that supply, and to protect the industry that provides it. It thus must be, simultaneously, a servant of the people and a servant of the private sector. Reviewing Secretary Glickman's speech in this light would prove instructive. That the U.S. Department of State would weigh in on this matter is especially revealing. Chapter 9 of the text does, in fact, talk about the new ally that environmentalists have found in the U.S. Department of State. Here, interestingly, the Department is taking a position at odds with most environmentalists. A useful discussion might ensue from questioning how the government positions affect the formation of policy, and whether this is a weakness in the regulatory protocol, especially given that, by and large, Americans have confidence in the protections afforded by government watchdogs.

The remaining documents are simply a small sample of the interest groups and individuals that will provide testimony over the months of pendency of the rule proposal, though I have made an effort to represent the full range of opinion. Again, it would be instructive to see what aspects of the proposed regulation draw the most fire and determine whether or not these are truly the most worrisome problems or simply the ones that will generate the most adverse public opinion. Then it would be interesting to compare these concerns with those presumably addressed by the government officials in the earlier documents.

Biotechnology

Release No. 0285.99

Remarks

As Prepared for Delivery
by

Secretary of Agriculture Dan Glickman
before the National Press Club on
New Crops, New Century, New Challenges:
How Will Scientists, Farmers,
And Consumers Learn to Love Biotechnology
And What Happens If They Don't?
Washington, D.C. - July 13, 1999

"Good afternoon. Thank you for coming.

"Let's think about this hypothetical situation for a moment: Let's suppose that today's salad was made with the new carrot from Press Club Farms, Inc. Farmers grow the new carrot on fewer acres because it yields more, and it's less expensive because it does not require any fertilizers or pesticides and can be harvested totally mechanically. In addition, it has more vitamin A & C than traditional varieties and stays crisper longer and keeps its fresh taste longer.

"But, because this carrot does not require as much labor, the farmers have had to lay off hundreds of employees. While it does not require any chemicals to flourish, this new carrot does affect the environment by making it difficult for other crops or plants in close proximity to survive. And though it's cheaper to begin with, it's only available from one company, which could result in a considerable premium over regular carrot seed.

"And what's the secret to this hypothetical new carrot? It's the latest advance from biotechnology -- produced with a gene from kudzu, an invasive weed.

"Sound far-fetched? It probably shouldn't: Remember the flavor-saver tomato? How many of you have heard of the so-called terminator gene which can keep a plant from reproducing? Today, nearly half the soybeans in the U.S. the stuff that is crushed and made into salad and cooking oil and that feeds most of the livestock we grow are produced from a variety that increases the plant's resistance to certain pesticides. Genetically-engineered corn with certain pest resistant characteristics is also rapidly displacing more traditional varieties. And, it gets even more interesting when you consider that researchers are looking at genetically-modified mosquitoes that cannot carry malaria.

"So, what do we think about this new carrot? Are we concerned about the environmental effects we still don't fully understand? What about the farm workers who are now unemployed? Should one company have a monopoly on it? And finally, are you concerned about these issues and about how it is produced? Would you still have eaten it if you knew about the kudzu gene? Should you have been told? Would you buy it?

"Folks, this is the tip of the biotechnology iceberg. There are many more questions that haven't yet been thought of, much less answered. But first of all, and if you come away with a dominant point from my remarks, it is that I want you to know that biotechnology has enormous potential.

"Biotechnology is already transforming medicine as we know it. Pharmaceuticals such as human insulin for diabetes, interferon and other cancer medications, antibiotics and vaccines are all products of genetic engineering. Just yesterday I read that scientists at Virginia Polytechnic Institute will process drugs from milk from genetically altered cows. One new drug has the potential to save hemophiliacs from bleeding to death. Scientists are also looking at bananas that may one day deliver vaccines to children in developing countries.

"Agricultural biotechnology has enormous potential to help combat hunger. Genetically modified plants have the potential to resist killer weeds that are, literally, starving people in Africa and other parts of the developing world.

"Biotechnology can help us solve some of the most vexing environmental problems: It could reduce pesticide use, increase yields, improve nutritional content, and use less water. We're employing bioengineered fungi to remove ink from pulp in a more environmentally sensitive manner.

"But, as with any new technology, the road is not always smooth. Right now, in some parts of the world there is great consumer resistance and great cynicism toward biotechnology. In Europe protesters have torn up test plots of biotechnology-derived crops and some of the major food companies in Europe have stopped using GMOs genetically-modified organisms in their products.

"Yesterday's news was that the WTO affirmed our view that the EU is unjustifiably blocking US ranchers from selling beef produced with completely tested and safe growth hormones. Today we're seeing that the G-8 agreed to a new review of food safety issues and, having myself just come back from France a couple of weeks ago, I can assure you that trade in GMOs is looming larger over US-EU trade relations in all areas.

"Now, more than ever, with these technologies in their relative infancy, I think it's important that, as we encourage the development of these new food production systems, we cannot blindly embrace their benefits. We have to ensure public confidence in general, consumer confidence in particular, and assure farmers the knowledge that they will benefit. "The important question is not, do we accept the changes the biotechnology revolution can bring, but are we willing to heed the lessons of the past in helping us to harness this burgeoning technology. The promise and potential are enormous, but so too are the questions many of which are completely legitimate. Today, on the threshold of this revolution, we have to grapple with and satisfy those questions so we can in fact fulfill biotechnology's awesome potential.

"To that end, today I am laying out 5 principles I believe should guide us in our approach to biotechnology in the 21st century. They are:

1. An Arm's Length Regulatory Process. Government regulators must continue to stay an arm's length, dispassionate distance from the companies developing and promoting these products; and continue to protect public health, safety and the environment.

2. Consumer Acceptance. Consumer acceptance is fundamentally based on an arm's length regulatory process. There may be a role for information labeling, but fundamental questions to acceptance will depend on sound regulation.

3. Fairness to Farmers. Biotechnology has to result in greater, not fewer options for farmers. The industry has to develop products that show real, meaningful results for farmers, particularly small and medium size family farmers.

4. Corporate Citizenship. In addition to their desire for profit, biotechnology companies must also understand and respect the role of the arm's length regulator, the farmer, and the consumer.

5. Free and Open Trade. We cannot let others hide behind unfounded, unwarranted scientific claims to block commerce in agriculture.

Arm's Length Regulatory Process

"When I was a school board member in Wichita, Kansas, one of my tasks was to study the level of student participation in the school lunch program. I quickly learned if the food didn't taste or look good, no matter how nutritious it was, the kids wouldn't eat it.

"With all that biotechnology has to offer, it is nothing if it's not accepted. This boils down to a matter of trust trust in the science behind the process, but particularly trust in the regulatory process that ensures thorough review -- including complete and open public involvement. The process must stay at arm's length from any entity that has a vested interest in the outcome.

"By and large the American people have trust and confidence in the food safety efforts of USDA, the FDA, EPA, CDC and others because these agencies are competent and independent from the industries they regulate, and are viewed as such, That kind of independence and confidence will be required as we deal with biotechnology.

"The US regulatory path for testing and commercializing biotechnology products as they move from lab to field to marketplace is over a decade old. We base decisions on rigorous analysis and sound scientific principles. Three federal agencies USDA, FDA, and EPA each play a role in determining the use of biotechnology products in the United States: USDA evaluates products for potential risk to other plants and animals. FDA reviews biotechnology's effect on food safety. And the EPA examines any products that can be classified as pesticides.

"Right now, there are about 50 genetically altered plant varieties approved by USDA. And so far, thanks to the hard work and dedication of our scientists, the system is keeping pace. But, as I said, the system is tried and tested, but not perfect and not inviolate and should be improved where and when possible.

"To meet the future demand of the thousands of products in the pipeline will require even greater resources, and a more unified approach and broader coordination.

"When I chaired the US delegation to the World Food Conference in Rome in 1996, I got pelted with genetically modified soybeans by naked protesters. I began to realize the level of opposition and distrust in parts of Europe to biotechnology for products currently on the market or in the pipeline.

"I believe that distrust is scientifically unfounded. It comes in part from the lack of faith in the EU to assure the safety of their food. They have no independent regulatory agencies like the FDA, USDA or EPA. They've had many food scares in recent years -- mad-cow disease, and in just the last several weeks, dioxin-tainted chicken -- that have contributed to a wariness of any food that is not produced in a traditional manner notwithstanding what the science says. Ironically they do not share that fear as it relates to genetically modified pharmaceuticals.

"But, GMO foods evoke in many circles a very volatile reaction. And that has created a serious problem for the U.S. and other countries as we try to sell our commodities in international markets.

"We need to make sure our regulatory system has the foresight to begin addressing issues even before they arise. So to keep pace with the accelerating growth of agricultural biotechnology, I am taking several additional steps to ensure we are fully prepared to meet the regulatory challenges of this new technology.

"Today I'm announcing that I will be asking for an independent scientific review of USDA's biotech approval process. The purpose of this review will be to ensure that, as we are faced with increasingly complex issues surrounding biotechnology, our scientists have the best information and tools to ensure our regulatory capabilities continue to evolve along with advances in the new technology. And to address complex issues like pharmaceutical producing plants or genetically modified livestock we will need to consult the experts, many of whom are outside USDA.

"Two of the more significant challenges we face are grower and consumer awareness, and improving monitoring on a long term basis. We do not have evidence the heavily publicized Monarch butterfly lab study appears to be happening in the field. But, the resulting attention to the reports and ensuing debate underscore the need to develop a comprehensive approach to evaluating long-term and secondary effects of biotech products.

"So, USDA will propose the establishment of regional centers around the country to evaluate biotech products over a long period of time and to provide information on an ongoing basis to growers, consumers, researchers and regulators.

"To strengthen biotechnology guidelines to ensure we can stay on top of any unforeseen adverse effects after initial market approval, I am requesting all developers of biotech products to report any unexpected or potentially adverse effects to the Department of Agriculture immediately upon discovery.

"Finally, we need to ensure that our regulators just regulate and only regulate. A few years ago, we created a food safety agency separate and distinct from any and all marketing functions to ensure that no commercial interests have even the appearance of influence on our decisions regarding food safety. It needs to be the same with biotechnology. The scientists who evaluate and approve biotech products for the market must be free of any hint of influence from trade support and other non-regulatory areas within USDA.

"We at USDA will undertake a review to reinforce the clear line between our regulatory functions and those that promote and support trade. This reaffirms our basic principle that we will remain scrupulously rigid in maintaining an arm's length regulatory process.

Consumer Acceptance

"However strong our regulatory process is, it is of no use if consumer confidence is low and if consumers cannot identify a direct benefit to them.

"I have felt for some time that when biotechnology products from agriculture hit the market with attributes that, let's say, reduce cholesterol, increase disease resistance, grow hair, lower pesticide and herbicide use, and are truly recognized as products that create more specific public benefits, consumer acceptance will rise dramatically.

"There's been a lot of discussion as to whether we should label GMO products. There are clearly trade and domestic implications to labeling to be considered in this regard. I know many of us in this room are sorting out these issues. At the end of the day many observers, including me, believe some type of informational labeling is likely to happen. But, I do believe that it is imperative that such labeling does not undermine trade and this promising new technology.

"The concept of labeling particular products for marketing purposes is not a radical one. For example, USDA has already decided that for a product to be certified as organic under our pending organic agriculture rules, a GMO product would not qualify. And that does not mean that USDA believes organic is safer or better than non-organic all approved foods are safe it just means that consumers are given this informed choice.

"There clearly needs to be a strong public education effort to show consumers the benefits of these products and why they are safe. Not only will this be the responsibility of private industry and government, but I think the media will play a vital role. It's important that the media treat this subject responsibly and not sensationalize or fan consumer fears. That's what we're seeing happen in the EU and the outcome is fear, doubt and outright opposition.

"What we cannot do is take consumers for granted. I cannot stress that enough. A sort of if-you-grow-it-they-will-come mentality. I believe farmers and consumers will eventually come to see the economic, environmental, and health benefits of biotechnology products, particularly if the industry reaches out and becomes more consumer accessible.

"But, to build consumer confidence, it is just like it is with the way we regulate our airlines, our banks and the safety of our food supply consumers must have trust in the regulatory process. That trust is built on openness. Federal agencies have nothing to hide. We work on behalf of the public interest. Understanding that will go a long way to solving the budding controversy over labeling and ensuring that consumers will have the ability to make informed choices.

Fairness to Farmers

"Like consumers, farmers need to have adequate choices made available to them. But today, American agriculture is at a crossroads. Farmers are currently facing extremely low commodity prices and are rightfully asking what will agriculture look like in the years to come and what will their roles be.

"That also means they have more responsibility and more pressure. And much of the pressure they face originates from sources beyond their control. We are seeing social and economic trends that have a powerful effect on how farmers do business. We are seeing increased market concentration, a rise in contracting, rapidly evolving technologies such as information power and precision agriculture in addition to biotechnology. We are seeing different marketing techniques such as organics, direct marketing, coops and niche markets, and an expansion of non-agricultural industrial uses for plants.

"One of my biggest concerns is what biotechnology has in store for family farmers. Consolidation, industrialization and proprietary research can create pitfalls for farmers. It threatens to make them servants to bigger masters, rather than masters of their own domains. In biotechnology, we're already seeing a heated argument over who owns what. Companies are suing companies over patent rights even as they merge. Farmers have been pitted against their neighbors in efforts to protect corporate intellectual property rights.

"We need to ensure that biotechnology becomes a tool that results in greater -- not fewer -- options for farmers. For example, we're already hearing concerns from some farmers that to get some of the more highly desirable non-GMO traits developed over the years, they might have to buy biotechnology seeds. For some, that's like buying the car of your dreams but only if you get it in yellow. On the other hand, stress-tolerant plants are in the pipeline which could expand agricultural possibilities on marginal lands which could be a powerful benefit to poor farmers.

"The ability of farmers to compete on a level playing field with adequate choices available to them and without undue influence or impediments to fair competition must be preserved. As this technology develops, we must achieve a balance between fairness to farmers and corporate returns.

"We need to examine all of our laws and policies to ensure that, in the rush to bring biotech products to market, small and medium family farmers are not simply plowed under. We will need to integrate issues like privatization of genetic resources, patent holders rights and public research to see if our approach is helping or harming the public good and family farmers.

"It is not the government who harnesses the power of the airwaves, but it is the government who regulates it. That same principle might come to apply to discoveries in nature as well. And that debate is just getting started.

Corporate Citizenship

"If the promises hold true, biotechnology will bring revolutionary benefits to society. But that very promise means that industry needs to be guided by a broader map and not just a compass pointing toward the bottom line.

"Product development to date has enabled those who oppose this technology to claim that all the talk about feeding the world is simply cover for corporate profit-making. To succeed in the long term, industry needs to act with greater sensitivity and foresight.

"In addition, private sector research should also include the public interest, with partnerships and cooperation with non-governmental organizations here and in the developing world ensuring that the fruits of this technology address the most compelling needs like hunger and food security. "Biotechnology developers must keep farmers informed of the latest trends, not just in research but in the marketplace as well. Contracts with farmers need to be fair and not result in a system that reduces farmers to mere serfs on the land or create an atmosphere of mistrust among farmers or between farmers and companies.

"Companies need to continue to monitor products, after they've gone to market, for potential danger to the environment and maintain open and comprehensive disclosure of their findings.

"We don't know what biotechnology has in store for us in the future, good and bad, but if we stay on top of developments, we're going to make sure that biotechnology serves society, not the other way around.

"These basic principles of good corporate citizenship really just amount to good long-term business practices. As in every other sector of the economy, we expect responsible corporate citizenship and a fair return. For the American people, that is the bottom line.

Free and Open Trade

"The issues I have raised have profound consequences in world trade. Right now, we are fighting the battles on ensuring access to our products on many fronts. We are not alone in these battles Canada, Australia, Mexico, many Latin American, African and Asian nations, agree with us that sound science ought to establish whether biotech products are safe and can move in international commerce.

"These are not academic problems. For 1998 crops 44% of our soybeans and 36% of our corn are produced from genetically modified seeds. While only a few varieties of GMO products have been approved for sale and use in Europe, many more have been put on hold by a de facto European moratorium on new GMO products.

"Two weeks ago I went to France and met with the French Agriculture Minister at the request of the US ambassador there, Felix Rohatyn, to see if we can break this logjam which directly threatens US-EU relations at a delicate time when we are commencing the next WTO round in Seattle.

"Quite frankly the food safety and regulatory regimes in Europe are so split and divided among the different countries that I am extremely concerned that failure to work out these biotech issues in a sensible way could do deep damage to our next trade round and effect both agricultural and non-agricultural issues. For that reason, the French Minister's agreement to have a short-term working group with USDA on biotech approval issues, and his willingness to come to the US in the fall to further discuss the situation, is encouraging.

"To forestall a major US-EU trade conflict, both sides of the Atlantic must tone down the rhetoric, roll up our sleeves and work toward conflict resolution based on open trade, sound science and consumer involvement. I think this can be done if the will is there.

"However, I should warn our friends across the Atlantic that, if these issues cannot be resolved in this manner, we will vigorously fight for our legitimate rights.

Conclusion

"Finally, I've established a Secretary's Advisory Committee on Agricultural Biotechnology -- a cross-section of 25 individuals from government, academia, production agriculture, agribusiness, ethicists, environmental and consumer groups. The committee, which will hold its first meeting in the fall, will provide me with advice on a broad range of issues relating to agricultural biotechnology and on maintaining a flexible policy that evolves as biotechnology evolves.

"Public policy must lead in this area and not merely react. Industry and government cannot engage in hedging or double talking as problems develop, which no doubt they will.

"At the same time, science will march forward, and especially in agriculture, that science can help to create a world where no one needs to go hungry, where developing nations can become more food self-sufficient and thereby become freer and more democratic, where the environmental challenges and clean water, clean air, global warming and climate change, must be met with sound and modern science and that will involve biotechnological solutions.

"Notwithstanding my concerns raised here today, I would caution those who would be too cautious in pursuing the future. As President Kennedy said, "We should not let our fears hold us back from pursuing our hopes."

"So let us continue to move forward thoughtfully with biotechnology in agriculture but with a measured sense of what it is and what it can be. We will then avoid relegating this promising new technology to the pile of what-might-have-beens, and instead realize its potential as one of the tools that will help us feed the growing world population in a sustainable manner.

"Thank you."

#

---

http://www.usda.gov/news/release/1999/07/0285

FDA Consumer magazine
January-February 2000

U.S. Food and Drug Administration

Are Bioengineered Foods Safe?

by Larry Thompson

Since 1994, a growing number of foods developed using the tools of the science of biotechnology have come onto both the domestic and international markets. With these products has come controversy, primarily in Europe where some question whether these foods are as safe as foods that have been developed using the more conventional approach of hybridization.

Ever since the latter part of the 19th century, when Gregor Mendel discovered that characteristics in pea plants could be inherited, scientists have been improving plants by changing their genetic makeup. Typically, this was done through hybridization in which two related plants were cross-fertilized and the resulting offspring had characteristics of both parent plants. Breeders then selected and reproduced the offspring that had the desired traits.

Today, to change a plant's traits, scientists are able to use the tools of modern biotechnology to insert a single gene--or, often, two or three genes--into the crop to give it new, advantageous characteristics. (See "Methods for Genetically Engineering a Plant.") Most genetic modifications make it easier to grow the crop. About half of the American soybean crop planted in 1999, for example, carries a gene that makes it resistant to an herbicide used to control weeds. About a quarter of U.S. corn planted in 1999 contains a gene that produces a protein toxic to certain caterpillars, eliminating the need for certain conventional pesticides.

In 1992, the Food and Drug Administration published a policy explaining how existing legal requirements for food safety apply to products developed using the tools of biotechnology. It is the agency's responsibility to ensure the safety of all foods on the market that come from crops, including bioengineered plants, through a science-based decision-making process. This process often includes public comment from consumers, outside experts and industry. FDA established, in 1994, a consultation process that helps ensure that foods developed using biotechnology methods meet the applicable safety standards. Over the last five years, companies have used the consultation process more than 40 times as they moved to introduce genetically altered plants into the U.S. market.

Although the agency has no evidence that the policy and procedure do not adequately protect the public health, there have been concerns voiced regarding FDA's policy on these foods. To understand the agency's role in ensuring the safety of these products, FDA Consumer sat down with Commissioner Jane E. Henney, M.D., to discuss the issues raised by bioengineered foods:

FDA Consumer: *Dr. Henney, what does it mean to say that a food crop is bioengineered?*

Dr. Henney: When most people talk about bioengineered foods, they are referring to crops produced by utilizing the modern techniques of biotechnology. But really, if you think about it, all crops have been genetically modified through traditional plant breeding for more than a hundred years.

Since Mendel, plant breeders have modified the genetic material of crops by selecting plants that arise through natural or, sometimes, induced changes. Gardeners and farmers and, at times, industrial plant breeders have crossbred plants with the intention of creating a prettier flower, a hardier or more productive crop. These conventional techniques are often imprecise because they shuffle thousands of genes in the offspring, causing them to have some of the characteristics of each parent plant. Gardeners or breeders then look for the plants with the most desirable new trait.

With the tools developed from biotechnology, a gene can be inserted into a plant to give it a specific new characteristic instead of mixing all of the genes from two plants and seeing what comes out. Once in the plant, the new gene does what all genes do: It directs the production of a specific protein that makes the plant uniquely different.

This technology provides much more control over, and precision to, what characteristic breeders give to a new plant. It also allows the changes to be made much faster than ever before.

No matter how a new crop is created--using traditional methods or biotechnology tools--breeders are required by our colleagues at the U.S. Department of Agriculture to conduct field testing for several seasons to make sure only desirable changes have been made. They must check to make sure the plant looks right, grows right, and produces food that tastes right. They also must perform analytical tests to see whether the levels of nutrients have changed and whether the food is still safe to eat.

As we have evaluated the results of the seeds or crops created using biotechnology techniques, we have seen no evidence that the bioengineered foods now on the market pose any human health concerns or that they are in any way less safe than crops produced through traditional breeding.

FDA Consumer: *What kinds of genes do plant breeders try to put in crop plants?*

Dr. Henney: Plant researchers look for genes that will benefit the farmer, the food processor, or the consumer. So far, most of the changes have helped the farmer. For example, scientists have inserted into corn a gene from the bacterium Bacillus thurigiensis, usually referred to as BT. The gene makes a protein lethal to certain caterpillars that destroy corn plants. This form of insect control has two advantages: It reduces the need for chemical pesticides, and the BT protein, which is present in the plant in very low concentrations, has no effect on humans.

Another common strategy is inserting a gene that makes the plant resistant to a particular herbicide. The herbicide normally poisons an enzyme essential for plant survival. Other forms of this normal plant enzyme have been identified that are unaffected by the herbicide. Putting the gene for this resistant form of the enzyme into the plant protects it from the herbicide. That allows farmers to treat a field with the herbicide to kill the weeds without harming the crop.

The new form of the enzyme poses no food safety issues because it is virtually identical to nontoxic enzymes naturally present in the plant. In addition, the resistant enzyme is present at very low levels and it is as easily digested as the normal plant enzyme.

Modifications have also been made to canola and soybean plants to produce oils with a different fatty acid composition so they can be used in new food processing systems. Researchers are working diligently to develop crops with enhanced nutritional properties.

FDA Consumer: *Do the new genes, or the proteins they make, have any effect on the people eating them?*

Dr. Henney: No, it doesn't appear so. All of the proteins that have been placed into foods through the tools of biotechnology that are on the market are nontoxic, rapidly digestible, and do not have the characteristics of proteins known to cause allergies.

As for the genes, the chemical that encodes genetic information is called DNA. DNA is present in all foods and its ingestion is not associated with human illness. Some have noted that sticking a new piece of DNA into the plant's chromosome can disrupt the function of other genes, crippling the plant's growth or altering the level of nutrients or toxins. These kinds of effects can happen with any type of plant breeding--traditional or biotech. That's why breeders do extensive field-testing. If the plant looks normal and grows normally, if the food tastes right and has the expected levels of nutrients and toxins, and if the new protein put into food has been shown to be safe, then there are no safety issues.

FDA Consumer: *You mentioned allergies. Certain proteins can cause allergies, and the genes being put in these plants may carry the code for new proteins not normally consumed in the diet. Can these foods cause allergic reactions because of the genetic modifications?*

Dr. Henney: I understand why people are concerned about food allergies. If one is allergic to a food, it needs to be rigorously avoided. Further, we don't want to create new allergy problems with food developed from either traditional or biotech means. It is important to know that bioengineering does not make a food inherently different from conventionally produced food. And the technology doesn't make the food more likely to cause allergies.

Fortunately, we know a lot about the foods that do trigger allergic reactions. About 90 percent of all food allergies in the United States are caused by cow's milk, eggs, fish and shellfish, tree nuts, wheat, and legumes, especially peanuts and soybeans.

To be cautious, FDA has specifically focused on allergy issues. Under the law and FDA's biotech food policy, companies must tell consumers on the food label when a product includes a gene from one of the common allergy-causing foods unless it can show that the protein produced by the added gene does not make the food cause allergies.

We recommend that companies analyze the proteins they introduce to see if these proteins possess properties indicating that the proteins might be allergens. So far, none of the new proteins in foods evaluated through the FDA consultation process have caused allergies. Because proteins resulting from biotechnology and now on the market are sensitive to heat, acid and enzymatic digestion, are present in very low levels in the food, and do not have structural similarities to known allergens, we have no scientific evidence to indicate that any of the new proteins introduced into food by biotechnology will cause allergies.

FDA Consumer: *Let me ask you one more scientific question. I understand that it is common for scientists to use antibiotic resistance marker genes in the process of bioengineering. Are you concerned that their use in food crops will lead to an increase in antibiotic resistance in germs that infect people?*

Dr. Henney: Antibiotic resistance is a serious public health issue, but that problem is currently and primarily caused by the overuse or misuse of antibiotics. We have carefully considered whether the use of antibiotic resistance marker genes in crops could pose a public health concern and have found no evidence that it does.

I'm confident of this for several reasons. First, there is little if any transfer of genes from plants to bacteria. Bacteria pick up resistance genes from other bacteria, and they do it easily and often. The potential risk of transfer from plants to bacteria is substantially less than the risk of normal transfer between bacteria. Nevertheless, to be on the safe side, FDA has advised food developers to avoid using marker genes that encode resistance to clinically important antibiotics.

FDA Consumer: *You've mentioned FDA's consultative process a couple of times. Could you explain how genetically engineered foods are regulated in the United States?*

Dr. Henney: Bioengineered foods actually are regulated by three federal agencies: FDA, the Environmental Protection Agency, and the U.S. Department of Agriculture. FDA is responsible for the safety and labeling of all foods and animal feeds derived from crops, including biotech plants. EPA regulates pesticides, so the BT used to keep caterpillars from eating the corn would fall under its jurisdiction. USDA's Animal and Plant Health Inspection Service oversees the agricultural environmental safety of planting and field testing genetically engineered plants.

Let me talk about FDA's role. Under the federal Food, Drug, and Cosmetic Act, companies have a legal obligation to ensure that any food they sell meets the safety standards of the law. This applies equally to conventional food and bioengineered food. If a food does not meet the safety standard, FDA has the authority to take it off the market.

In the specific case of foods developed utilizing the tools of biotechnology, FDA set up a consultation process to help companies meet the requirements. While consultation is voluntary, the legal requirements that the foods have to meet are not. To the best of our knowledge, all bioengineered foods on the market have gone through FDA's process before they have been marketed.

Here's how it works. Companies send us documents summarizing the information and data they have generated to demonstrate that a bioengineered food is as safe as the conventional food. The documents describe the genes they use: whether they are from a commonly allergenic plant, the characteristics of the proteins made by the genes, their biological function, and how much of them will be found in the food. They tell us whether the new food contains the expected levels of nutrients or toxins and any other information about the safety and use of the product.

FDA scientists review the information and generally raise questions. It takes several months to complete the consultation, which is why companies usually start a dialog with the agency scientists nearly a year or more before they submit the data. At the conclusion of the consultation, if we are satisfied with what we have learned about the food, we provide the company with a letter stating that they have completed the consultation process and we have no further questions at that time.

FDA Consumer: *Since genes are being added to the plant, why doesn't FDA review biotech products under the same food additive regulations that it reviews food colors and preservatives?*

Dr. Henney: The food additive provision of the law ensures that a substance with an unknown safety profile is not added to food without the manufacturer proving to the government that the additive is safe. This intense review, however, is not required under the law when a substance is generally recognized as safe (GRAS) by qualified experts. A substance's safety can be established by long history of use in food or when the nature of the substance and the information generally available to scientists about it is such that it doesn't raise significant safety issues.

In the case of bioengineered foods, we are talking about adding some DNA to the plant that directs the production of a specific protein. DNA already is present in all foods and is presumed to be GRAS. As I described before, adding an extra bit of DNA does not raise any food safety issues.

As for the resulting proteins, they too are generally digested and metabolized and don't raise the kinds of food safety questions as are raised by novel chemicals in the diet. The proteins introduced into plants so far either have been pesticides or enzymes. The pesticide proteins, such as BT, would actually be regulated by EPA and go through its approval process before going on the market. The enzymes have been considered to be GRAS, so they have not gone through the food additive petition process. FDA's consultation process aids companies in determining whether the protein they want to add to a food is generally recognized as safe. If FDA has concerns about the safety of the food, the product would have to go through the full food additive premarket approval process.

FDA Consumer: *Why doesn't FDA require companies to tell consumers on the label that a food is bioengineered?*

Dr. Henney: Traditional and bioengineered foods are all subject to the same labeling requirements. All labeling for a food product must be truthful and not misleading. If a bioengineered food is significantly different from its conventional counterpart--if the nutritional value changes or it causes allergies--it must be labeled to indicate that difference. For example, genetic modifications in varieties of soybeans and canola changed the fatty acid composition in the oils of those plants. Foods using those oils must be labeled, including using a new standard name that indicates the bioengineered oil's difference from conventional soy and canola oils. If a food had a new allergy-causing protein introduced into it, the label would have to state that it contained the allergen.

We are not aware of any information that foods developed through genetic engineering differ as a class in quality, safety, or any other attribute from foods developed through conventional means. That's why there has been no requirement to add a special label saying that they are bioengineered. Companies are free to include in the labeling of a bioengineered product any statement as long as the labeling is truthful and not misleading. Obviously, a label that implies that a food is better than another because it was, or was not, bioengineered, would be misleading.

FDA Consumer: *Overall, are you satisfied that FDA's current system for regulating bioengineered foods is protecting the public health?*

Dr. Henney: Yes, I am convinced that the health of the American public is well protected by the current laws and procedures. I also recognize that this is a rapidly changing field, so FDA must stay on top of the science as biotechnology evolves and is used to make new kinds of modifications to foods. In addition, the agency is seeking public input about our policies and will continue to reach out to the public to help consumers understand the scientific issues and the agency's policies.

Not only must the food that Americans eat be safe, but consumers must have confidence in its safety, and confidence in the government's role in ensuring that safety. Policies that are grounded in science, that are developed through open and transparent processes, and that are implemented rigorously and communicated effectively are what have assured the consumers' confidence in an agency that has served this nation for nearly 100 years.

Larry Thompson is a member of FDA's public affairs staff.

---

http:www.fda.gov/fdac/features/2000/100-bio.html

Healthy Harvests: Growth Through Biotechnology

Fact sheet released by the U.S. Department of State, March 21, 2000

"Neither politics nor protectionism should deny the world's consumers the right to benefit from technological breakthroughs in the production of food."

—U.S. Secretary of State Madeleine Albright

- **Biotechnology is employed all around us in many everyday products, from the clothes we wear to the cheese we eat.** For centuries, farmers, ranchers, bakers and brewers have been using traditional techniques to make and modify plants and food products — wheat being an old example and the nectarine being a recent example. Today, biotechnology employs modern scientific techniques to allow us to improve or modify plants, animals and microorganisms with greater precision and predictability.

- **Consumers should have the opportunity to choose from the widest possible array of safe products.** Biotechnology can offer consumers a wide array of choices — not just in agriculture and food production, but also in medicine and fuel resources.

- **Responsible biotechnology can offer enormous potential benefits.** It is in the best interest of developing and developed countries to support additional research to help biotechnology achieve its full potential.

    - **Biotechnology helps the environment.** By allowing farmers to reduce the use of herbicides and pesticides, the first generation of biotechnology products is helping to reduce herbicide and pesticide use, and future products are expected to yield more environmental benefits. Reduced herbicide and pesticide use means a smaller risk of toxic contamination of both surface and groundwater. In addition, herbicides used in conjunction with genetically engineered plants are often safer for the environment than the herbicides they replace. Bioengineered crops also may reduce the need for farming practices, such as tillage, that result in soil loss.

    - **Biotechnology has tremendous potential to help fight hunger.** Biotechnology developments offer significant potential benefit for developing countries where almost a billion people live in poverty and suffer from chronic hunger. By increasing crop yields and making crops disease — and drought — resistant, biotechnology could reduce food shortages for a world population expected to exceed 8 billion by 2025 — an increase of more than 30 percent from our current population. Researchers are developing varieties of staple crops to enable them to survive harsh conditions such as droughts and floods.

- ◦ **Biotechnology helps combat diseases.** By developing and improving medicines, biotechnology helps fight diseases. Biotechnology has given us new medicinal tools to treat heart disease, multiple sclerosis, hemophilia, hepatitis, and AIDS, among other illnesses. Biotech foods are being developed that may enable the world's poor to receive low cost and widely available important vaccines and vitamins.
- ◦ **Biotechnology could yield significant benefits to health.** By boosting the nutritional value of foods, biotechnology can be used to improve the quality of basic diets. For example, rice and corn varieties with enriched protein contents are in development. In the future, consumers will be able to benefit from cooking oils with reduced saturated fat content derived from genetically engineered corn, soybean, or canola plants. In addition, genetic engineering may be used to produce foods with enhanced levels of vitamin A, which would help address a major cause of blindness in developing countries. Genetic engineering can also offer health benefits beyond nutrition, as techniques have been developed that can be used to remove specific allergenic proteins from food products or to delay spoilage.

- **Biotech products, reviewed by U.S. regulatory authorities, are safe.** In fact, information available indicates that biotech products currently on the market are as safe as conventional foodstuffs — both for human health and for the environment. U.S. regulatory authorities are constantly evaluating their procedures for ensuring the safety of biotech products. If there were scientific evidence that biotech food posed a threat to human health, such foods would not be on the market in the United States.

- **Blocking trade in safe agricultural biotech products reduces consumer choice, forces consumers to pay higher prices for basic products, and discourages research on future beneficial products.**

- **Sound science remains the best foundation for food safety and environmental decisions.** Legitimate concerns about possible environmental impacts must not be ignored. The United States is committed to an open dialogue, backed by science, with all interested parties in the biotechnology debate. At the same time, the public should not be denied the right to choose new products because of misinformation that has raised unfounded fears.

- **Accurate information regarding the safety of biotech products should be publicly available.** Transparency is central to increasing the level of public trust in sound science. The United States believes that it is important to respond to the public's concern about biotechnology and urges all countries to provide accurate information regarding the safety of these products.

---

http://www.state.gov/www/global/oes/fs-biotech_growth_000321.html

HHS *NEWS*

U.S. Department of Health and Human Services

P00-10
FOR IMMEDIATE RELEASE
May 3, 2000

FOOD AND DRUG ADMINISTRATION
Print Media: 301-827-6242
Broadcast Media: 301-827-3434
Consumer Inquiries: 888-INFO-FDA

FDA TO STRENGTHEN PRE-MARKET REVIEW OF BIOENGINEERED FOODS

The Food and Drug Administration (FDA) announced today plans to refine its regulatory approach regarding foods derived through the use of modern biotechnology. The initiatives unveiled stem in part from input received during FDA's public outreach meetings held late last year and build upon programs already underway at FDA to help ensure the safety of all foods.

"FDA's scientific review continues to show that all bioengineered foods sold here in the United States today are as safe as their non- bioengineered counterparts, "said Jane E. Henney, MD, Commissioner of Food and Drugs. "We believe our initiatives will provide the public with continued confidence in the safety of these foods."

FDA will publish a proposed rule mandating that developers of bioengineered foods and animal feeds notify the agency when they intend to market such products. FDA also will require that specific information be submitted to help determine whether the foods or animal feeds pose any potential safety, labeling or adulteration issues.

Although the current consultative process has worked well, and the agency believes it has been consulted on all bioengineered foods and feeds currently on the market, FDA will propose to strengthen this process by specifically requiring developers to notify the agency of their intent to market a food or animal feed from a bioengineered plant at least 120 days before marketing. After reviewing the company's submission, FDA will issue a letter to the firm describing its conclusion about the regulatory status of the food or animal feed. To make sure that consumers also have access to product information, FDA will propose that submitted information and the agency's conclusions be made available to the public, consistent with applicable disclosure laws, by posting them on the FDA Web site for easy viewing.

In a related step, the agency will augment its food and veterinary medicine advisory committees by adding scientists with agricultural biotechnology expertise. FDA will use these committees to address over- arching scientific questions pertaining to bio-engineered foods and animal feed.

FDA also announced today plans to draft labeling guidance to assist manufacturers who wish to voluntarily label their foods being made with or without the use of bio-engineered ingredients. The guidelines will help ensure that labeling is truthful and informative. To receive maximum consumer input, FDA will develop the guidelines with the use of focus groups and will seek public comment on the draft guidance.

---

http://www.fda.gov/bbs/topics/NEWS/NEW00726.html

Plant Biotechnology Regulation:

Science-Based and Consumer Accessible From Plow to Plate

(From the American Crop Protection Association)

Since the introduction of plant biotechnology on the American farm began in the mid-1990s, farmers have planted millions of acres of biotech corn, cotton, soybeans, fruit and other crops. While these new varieties have enhanced crop protection, reduced production costs and increased yields, they have raised concern among some consumers, legislators and activists. What is government's role in assuring human, animal and environmental safety? To what degree are biotech-derived crops reviewed and regulated?

The Regulation: Comprehensive, Coordinated and Transparent

The U.S. government has regulated biotechnology since its inception in the 1970s. For many years, responsibility rested with the National Institutes of Health (NIH), which developed the protocols used to evaluate the safety of biotechnology research in the laboratory.

But with the rapid evolution of plant biotechnology in the early 1980s, additional regulation was needed and the "Coordinated Framework for Regulation of Biotechnology" was adopted. The framework apportions regulatory responsibility to the U.S. Department of Agriculture (USDA), the Environmental Protection Agency (EPA), and the Food and Drug Administration (FDA). USDA, EPA, and FDA, working within the coordinated framework and existing laws, evaluate the safety of plant biotechnology products from inception to final approval.

The Regulators: USDA, EPA, and FDA

- *USDA*

  Under the Federal Plant Pest Act, USDA—primarily through its Animal and Plant Health Inspection Service (APHIS)—oversees field testing of biotech seeds and plants to assure their safe release and that no harm is done to the environment, particularly native plants and organisms.

- *EPA*

  Under the Federal Insecticide, Fungicide and Rodenticide Act (FIFRA) and the Federal Food, Drug and Cosmetic Act (FFDCA), EPA evaluates the pesticidal properties of biologically-enhanced plants—the traits of insect protected and virus resistance crops, for example. For herbicide resistant crops, EPA regulates the herbicide applied to the crop, not the resistance trait itself.

- *FDA*

  Under FFDCA, FDA assesses the safety of all foods and animal feeds, including those produced through biotechnology.

Although traditionally bred plants, such as heartier hybrids, generally do not need federal pre-market approval, plant biotechnology products are screened by at least one and often by as many as three federal agencies. For example, tomatoes developed with delayed ripening qualities would be reviewed by USDA for field safety and FDA for food safety. However, if a soybean or corn plant, for example, has enhanced protection against insects EPA would be involved in the regulatory process, along with USDA and FDA.

The Process: Rigorous, Science-Based and Open

From conception to commercial introduction, it can take up to 10 years to bring a biotech plant to market. Regulatory oversight is constant throughout the process. There are 10 separate points at which federal regulators can question and/or halt development of a biotech plant variety. Throughout, the public has ample opportunity for participation and comment, and data on which regulatory decisions are based are readily available.

Regulating Plant Biotechnology: 10 Steps to Safety; 6 Opportunities to Speak Out

NIH

Discovery of a potentially marketable biotech plant triggers NIH involvement.

1. *Biosafety Committee Review:*
   Following NIH guidelines, an advisory group of scientists, company employees and citizens evaluates the plant for potential health and environmental risks. If further laboratory research poses unacceptable risk, the committee may recommend a halt to development.

Public review and comment.

USDA

There are four points at which USDA may advance or halt development of a biotech plant product.

2. *Greenhouse Approval:*
   USDA determines the adequacy of research facilities, such as greenhouses, for biotech testing and development.

3. *Field Trial Authorization:*
   Plant biotech developers must receive USDA approval for field trials and submit summary reports. USDA-APHIS personnel and state agriculture officials may inspect the trials at any time to ensure the research is being conducted safely. Particular scrutiny is given to limit the risk of possible "outcrosses," the unintentional breeding of domestic plants with related wild species. In addition, most plant developers avoid imparting traits that could increase the competitiveness or other undesirable properties of "weedy" relatives.

4. *Authorization to Transport Seed:*
   USDA also oversees import and shipment of biotech seeds from the greenhouse or laboratory to the field trial site.

5. *Permission to Commercialize:*
   Before a biotech crop can be grown, tested or used for traditional plant breeding without further USDA action, APHIS technical experts review all field trial studies. A "determination of non-regulated status" then is granted only after APHIS concludes the plant will not become a pest nor pose any significant risk to the environment or to wildlife. The reviews, which can take 10 months or longer, require developers to provide detailed scientific data on:

   - Environmental Effects—whether the biotech plant variety could crossbreed with other plants, allowing them to "outcompete" with other plants in the ecosystem.
   - Wildlife Effects—whether the plant could have adverse effects on wildlife, including birds, beneficial insects or mammals.
   - Weediness—whether the plant could become an uncontrollable or "super" weed.

Prior to the APHIS permit, plant developers must submit copies of their applications to the state departments of agriculture for review. If at any time the plant is found to be a pest, APHIS can stop its further development and sale or allow use under restricted conditions.

Public is notified and comment is invited through publication in the Federal Register. For information, go to http://www.usda.gov/agencies/biotech/laws.html.

EPA

If a biotechnology-derived plant exhibits the capacity to protect itself as if it has been treated with a chemical pesticide, EPA takes three steps.

6. *Experimental Use Permit Approval:*
For tests of 10 acres or more, EPA must grant an experimental use permit (EUP).

Public is notified and comment is invited through publication in the Federal Register. For information, go to http://www.epa.gov/pesticides/biopesticides or to http://www.epa.gov/opptintr/biotech/index.html.

7. *Food Tolerance or Exemption:*
In establishing the limits (tolerances) for the amount of pest-control proteins in foods derived from the biotech plants, EPA examines:

- Product Characterization—Where in the plant are the new traits expressed? Do traits behave in the plant the same way they behave in nature? What is the mode of action or specific pesticidal substance produced in the plant?
- Toxicology—Is the protein produced by the introduced trait or gene toxic? How long does it take the protein to break down in gastric and intestinal fluids?
- Allergenicity—Does the protein degrade like other dietary proteins? Answering this question helps EPA determine whether the protein may trigger an allergic reaction, thereby requiring a label.
- Non-target Organisms—Is the protein toxic to birds, beneficial insects, fish or other organisms? If so, will these organisms be exposed to the protein? For organisms that are exposed, data must be generated to ensure the safety of the expressed protein(s).
- Environmental Fate—How fast does the pesticidal protein in the plant's tissue break down in the soil?
- Potential Pest Resistance—What steps need to be followed to manage potential pest resistance?

Public is notified and comment is invited through publication in the Federal Register.

8. *Product Registration:*
During this final 18-month step, EPA reviews all relevant environmental and toxicological studies before deciding to register the product. if at any time the biotech plant is found to be unsafe, EPA can stop its further development and sale.

The public is notified and comments requested.

FDA

FDA primarily oversees food and feed safety.

9. *Review Process:*
FDA considers foods and feeds derived from biotechnology to be as safe and nutritionally equivalent as their conventionally bred counterparts. Under a 1992 policy, the agency treats biotech foods just like conventionally produced foods, unless they contain ingredients or demonstrate attributes that are unusual for the product. Even though the process is voluntary, all food biotechnology products currently on the market have received FDA consultation prior to their market introduction. However, acknowledging the need to increase consumer confidence in the regulatory process, FDA has decided to require developers to notify it at least four months in advance of releasing any biotech ingredients for food and animal feed and to supply the agency with their research data for review.

FDA review currently includes a thorough safety assessment that compares every significant parameter of the biotech plant with its traditional counterpart:

- Assessment and Testing of Introduced Material—Is the inserted gene already present in another food source? Is it comparable to proteins already present in foods or is it a protein without a history of human consumption? Is it allergenic? Even when the history of the inserted material is well known, studies are conducted to ensure its safety and to identify any unexpected effects in the plant.

- Biological and Agronomic Characteristics—Are the biological and agronomic properties of the biotech plant different from the parental equivalent?

- Nutritional Composition—Studies are performed to determine whether nutrients, vitamins, and minerals in the new plant occur at the same level as in the conventionally bred plant.

Virtually all crop plants have been changed through traditional plant breeding. Thus, FDA requires labeling of food derived from plant varieties developed through biotechnology only when there are scientifically established issues of safety (such as introduction of a known allergen); a significant change in nutrients or composition (such as nutritionally significant higher or lower level of certain vitamins); or a change in identity, in which traditional names do not apply (such as broccoflower).

When no changes are detected, FDA concludes the biotech crop to be substantially equivalent to the conventional crop with respect to nutrition and food safety.

FDA also has authority to stop development and sale of a biotech product if at any time it is found to be unsafe.

Public is notified and comment is invited through publication in the Federal Register. For information, go to http://www.vm.cfsan.fda.gov/~lrd/biotechm.html.

10. *After Commercialization:*
All three regulatory agencies have the legal power to demand immediate removal from the marketplace of any product should new, valid data indicate a question of safety for consumers or the environment.

Take Action Now! *Home Page* *Other Facts and Issues*

The Food and Drug Administration's New Proposal on Genetically Engineered Foods
1st Draft Analysis

Joseph Mendelson, Legal Director
Center for Food Safety
17 January 2001

Introduction.

On January 18, 2001, the United States Food and Drug Administration (FDA) released its long awaited revisions to its policy on genetically engineered foods. These revisions consist of two documents: (1) A proposed regulation entitled "Premarket Notice Concerning Bioengineered Foods," and (2) a voluntary guidance entitled "Guidance for Industry: Voluntary Labeling Indicating Whether Foods Have or Have Not Been Developed Using Bioengineering." See, 66 Federal Register 4706, 66 Federal Register 4839 (January 18, 2001). What follows is a brief synopsis and analysis of the proposals. The Center for Food Safety will provide a more in-depth analysis in the coming days.

I. Proposed Pre-market Notification

The new FDA proposal reaffirms its 1992 Policy on genetically engineered foods and does not require any mandatory pre-market safety testing of genetically engineered foods. The FDA restates its position that genetically engineered foods are assumed generally recognized as safe so they are not subject to mandatory pre-market safety review under the Federal Food Drug and Cosmetic Act's Food additive petition process. Despite the filing of the CFS, et al. legal petition (March 19, 2000) and the public testimony of numerous people, the Agency maintains this position by asserting that the Agency is aware of no new science calling into question the safety of GE foods.

In addition to not requiring specific pre-market safety testing, the FDA maintains that companies may voluntarily consult with FDA concerning the safety of their foods. This appears to be a regression from its past suggestions that these consultations would be mandatory. In the proposal, FDA readily admits the voluntary nature of such consultation will continue:

> "However, because the consultation process is voluntary, food producers could choose not to notify FDA. Additionally, as food producers in countries that export foods to the United States begin to adopt bioengineered varieties, they may choose not to participate in the voluntary consultation process." 66 Fed. Reg. 4727.

The only really new requirement put forth in the proposed regulation is that a genetically engineered (GE) food producer must send a letter to FDA notifying FDA of its intent to market such a food. The letter must be sent 120 days in advance of marketing.

The letter (known as a pre-market biotechnology notice) must include information of several topics including descriptions of the foods, methods of the food's development, use of antibiotic resistant markers, information about substances introduced into the foods (including allergencity issues), and information comparing it to a conventional, comparable food.

As exposed in the StarLink incident, the FDA currently has no immediate ability to trace a GE food through the food supply should a harm to public health become apparent. As a result, the FDA specifically seeks comment as to whether a producer's pre-market notification letter should be required to include methods by which its food could be detected once in the marketplace.

There are several exemptions to the notification requirement. GE foods already on the market are exempt from the new requirements. CFS will have to analyze the other exemptions more closely.

II. Public Disclosure

The FDA trumpets that this new process will make safety and review information about GE foods more transparent and accessible to the public. However, this is not likely to be a reality for a number of reasons. First, the FDA asserts that pre-market notification letters and voluntary consultation documentation filed with the Agency will be subject to the provisions of the Freedom of Information Act (FOIA). While that may appear transparent, the FDA notes that producers of GE foods may claim that any such information (including the premarket notification letter itself) is a trade secret or confidential business information subject to an exemption to public disclosure requirements. It is also important to note that the continued voluntary nature of the consultation submissions may effectively prevent public scrutiny from the real safety and testing data of GE foods. Recent court cases have made it extremely difficult for the public to get FOIA release of documents voluntarily submitted to an agency and later claimed as trade secrets or confidential business information(CBI).

FDA has stated it will make available its response letters sent back to GE food producers using the notification procedures and any information not claimed by a producers as trade secrets or CBI.

III. Environmental Review

Despite formally adopting a regulation allowing genetically engineered foods to enter the commercial market unobstructed, the FDA has exempted its proposal from environmental review procedures under the National Environmental Policy Act.

IV. Voluntary Labeling Guidance

While admitting that the public comments it has received overwhelmingly support labeling, FDA reasserts that it will not require the mandatory labeling of GE foods. In concert with that decision, FDA has release non-binding guidance on how labeling should take place for producers who want to voluntarily label their products. The guidance document suggests that FDA will attempt to severely limit the type of voluntary labels claiming that products avoid using GE foods. (See attached memo "Why Voluntary Labeling of Genetically Engineered Food Won't Help Consumers"). In particular, the FDA makes the following assertions:

- Use of the terms "GMO free" and "GM free" will be misleading in most foods;
- Use of the term "free" in claim of absence of genetically engineered material will be misleading because consumers will assume an absolute zero level of GE contamination, and food producers cannot test to meet such assurances level and a threshold level defining "free" has not been established; and
- Any statement about not being genetically engineered implying that the labeled food is superior to foods not so labeled would be misleading.

It should be noted that in the case of genetically engineered bovine growth hormone (rBGH) small dairies were sued by Monsanto for rBGH-free labels because such products asserted superiority over dairy products derived for rBGH treated cows and impugned the safety of Monsanto's product. While the guidance is non-binding legally, the last bullet point appears to set up a scenario similar to rBGH for non-GE food voluntary labeling.

The FDA does say that cumbersome statements such as, "we do not use ingredients that were produced using biotechnology" would not be misleading.

The guidance also reaffirms that the voluntary labeling scheme shifts the financial burden onto those food companies not using GE foods by reasserting that companies using non-GE food labeling claims need verification systems in place including chain of custody documentation, testing results or be certified organic.

V. Public Comment Period

The good news is that these rules are not final. The FDA, as required by law, has opened a comment period to hear from the public. This comment period lasts until April 3, 2001.

Comments can be submitted by writing to: Docket No. 00N-1396 & Docket No. 00D-1598, FDA Commissioner, Dockets Management Branch (HFA-305), Food and Drug Administration, 5630 Fishers Lane, Room 1061, Rockville, MD 20852.

Comments can be submitted electronically through the Center for Food Safety's action website found at: http://www.foodsafetynow.org/

Conclusion.

The Center for Food Safety believes that the new FDA proposal is contrary to law and wholly inadequate. It does little, if anything, to provide the public with assurances that GE foods are safe for human health and the environment. It also fails to provide consumers their right-to-know by limiting the labeling of GE foods to a voluntary labeling scheme. The new proposal can be summed up as follows:

- Does not require mandatory pre-market safety testing;
- Does not require pre-market environmental review;
- Does not require mandatory labeling of GE foods;
- Limits voluntary non-GE labeling;
- Only requires a letter of notification prior to the marketing of a GE food; and
- Is unlikely to provide the public with adequate information on GE foods for independent review.

---

http://www.centerforfoodsafety.org/facts&issues/CFSNewFDAAnalysis.html?cam_id=70

THE ABC'S OF GE FOOD

The Science and the Regulations

Rachel Massey is a writer for the Environmental Research Foundation.

Editor's Note: This is the fourth and final article in a series reviewing the problems with genetically engineered crops. This article reviews the regulation of GE crops.

Biotechnology corporations want the world to believe that the U.S. government has fully tested genetically engineered crops for ecological and human health hazards. Even though three federal agencies — the Food and Drug Administration, the Department of Agriculture, and the Environmental Protection Agency — are responsible for regulating genetically engineered foods, there is no guarantee that those foods have been systematically tested for harmful effects. In the rush to promote genetic engineering, safety testing has fallen through the cracks.

Biotechnology corporations would also like us to believe that genetically engineered foods have been embraced by the public. In fact, genetically engineered foods are not labeled, so the public has no knowledge — and no choice — about purchasing and eating them.

U.S. Food and Drug Administration

The Food and Drug Administration (FDA) issued its basic policy statement on genetically engineered foods in 1992. Under this policy, the FDA considers genetically engineered foods to be "generally recognized as safe" (abbreviated GRAS), unless the manufacturer tells the FDA there is reason for concern. Foods considered GRAS are not subject to pre-market safety testing.

According to the FDA, the need for safety testing depends on the characteristics of a food, not on the methods used to produce it. In other words, the fact that a food was produced using genetic engineering is not sufficient to trigger safety tests.

The FDA's 1992 policy says that a genetically engineered food must be labeled if it "differs from its traditional counterpart such that the common or usual name no longer applies to the new food, or if a safety or usage issue exists to which consumers must be alerted." For example, a tomato containing peanut genes might need to be labeled so that people with peanut allergies could avoid it. But the FDA allows biotechnology corporations to decide whether a hazard of this sort exists. Under this policy, we can find out the fat, cholesterol, sodium, potassium, carbohydrate, and protein content of our breakfast cereal, but we can't find out whether it contains antibiotic-resistance genes, viral promoters, or proteins normally produced only by bacteria.

In 1998, a coalition of non-governmental organizations, scientists, and others filed a lawsuit against the FDA for failing to fulfill its regulatory duties. The suit forced the FDA to release internal documents showing the agency's own scientists had strongly opposed the 1992 policy.

The lawsuit also forced the FDA to release details of its safety evaluation of the first genetically engineered food that entered U.S. supermarkets, the Flavr Savr tomato. Calgene, the company that developed the Flavr Savr, submitted to FDA officials the results of three safety tests in which rats were fed either ordinary tomatoes or engineered tomatoes. After 28 days, researchers examined the rats' stomachs. The results were inconsistent. The first study showed no unusual effects. In the second study, some of the rats fed genetically engineered tomatoes developed gastric erosions (damage to the lining of the stomach). In the third study, gastric erosions appeared in some of the rats fed ordinary tomatoes, as well as in some of the rats fed engineered potatoes.

According to Calgene, the gastric erosions were unrelated to eating genetically engineered tomatoes, but the company was unable to explain what caused them. An FDA staff scientist who reviewed the data said there were "doubts as to the validity of any scientific conclusion(s) which may be drawn from the studies' findings" because Calgene could not explain the variations in the test results. In spite of the misgivings expressed by its own staff, the FDA categorized the Flavr Savr tomato as GRAS and approved it for sale. (The Flavr Savr did not sell well, so it has since disappeared from stores.)

> The FDA is more concerned about protecting trade secrets than protecting the public's right to know what is in its food.

In January 2001, the FDA proposed new regulations for genetically engineered food. In its proposal, the FDA said it does not know of "any new scientific information that raises questions about the safety of bioengineered food currently being marketed," and stated once again that genetically engineered foods are "generally recognized as safe." To make this claim, the FDA had to ignore scientific information explicitly brought to its attention during the previous year. For example, the nonprofit Center for Food Safety and partner organizations filed a legal petition in March 2000 asking the FDA to require safety testing, environmental impact assessments, and labeling for all genetically engineered foods. The petition contained a thorough review of new scientific evidence on safety concerns associated with these foods.

The proposed regulations still fail to require either pre-market safety testing or labeling. Instead, the FDA's new proposed regulations would simply require producers to notify the agency 120 days before bringing a new genetically engineered food to market. This "pre-market biotechnology notice" would include information such as whether a product contains antibiotic-resistance marker genes and whether it is likely to produce allergic reactions. The FDA says it will make a list of these notices available to the public, but that the list may not be complete. In some cases, the FDA says, the existence of a pre-market biotechnology notice could be considered "confidential commercial information." As a result, under the proposed regulations a company could still market a genetically engineered food without notifying the public. In other words, the FDA is more concerned about protecting trade secrets than protecting the public's right to know what is in its food. The proposed regulations are open for public comment until May 3, 2001.

The FDA has also proposed to institute non-binding guidance for voluntary labeling. This guidance is clearly not intended for companies using genetically engineered crops, which have nothing to gain by telling consumers what is in their food. Instead, the guidance undermines consumers' right to know what they are buying and threatens to limit the free speech of organic and other food producers, by discouraging the use of labels with phrases such as "biotech free" or "no genetically engineered materials." The FDA says these labels would be misleading because ordinary foods could be contaminated with the products of genetic engineering. The agency also says these phrases could imply that non-engineered food is superior to engineered food, which, the FDA says, would also be misleading.

U.S. Department of Agriculture

The U.S. Department of Agriculture (USDA) is responsible for regulating "plant pests" — organisms that could cause harm to plants. According to the USDA, genetically engineered plants may be plant pests if they contain genetic material from organisms, such as some bacteria, included on an official plant pests list. Plants engineered with genes from an organism not on the list may escape USDA regulation entirely. Even when genes from a listed plant pest are involved, it is up to the manufacturer to decide whether or not to report the engineered plant to the USDA as a pest.

Under USDA rules, genetically engineered crops that are considered plant pests must be approved for field testing before they are approved for commercial planting. After conducting field tests, the crop developer can apply for "nonregulated status." With that status, a crop can be planted commercially without further oversight from the USDA. It is up to the developer to decide what data to submit with its application for nonregulated status. According to a recent article in *American Scientist*, many tests that companies submit to the USDA are poorly designed, so that even if adverse ecological effects occur, the tests may not reveal them.

U.S. Environmental Protection Agency

Crops can be engineered to kill certain insects by adding a gene derived from the bacterium *Bacillus thuringiensis* (*Bt*). Under its authority to regulate pesticides, the Environmental Protection Agency (EPA) is responsible for evaluating the health and environmental consequences of these engineered plants, which are themselves pesticidal.

The EPA has registered pesticidal crops for five years, but it does not have a standard testing system tailored to the hazards posed by genetically engineered crops. The EPA says it is reviewing existing registrations for *Bt* corn and cotton this year to decide whether it is safe to continue growing them. Of course, it would have been better to conduct a thorough review of likely environmental effects before, rather than after, allowing these crops to be planted on a large scale.

When the EPA registers a chemical pesticide for use on food crops, it establishes a tolerance level — an amount of pesticide residue that is allowable on food. However, the EPA has exempted *Bt* pesticides engineered into crops from this requirement.

As we discussed in Engineering the Landscape, pesticidal crops can promote the development of *Bt*-resistant pest populations. Despite ample scientific knowledge about this danger, the EPA waited until December 1999 to issue requirements for resistance management. These requirements are designed to slow, not stop, the development of resistance.

In the past five years, corporations have introduced a powerful new technology into our food system without understanding the basics of how it works. We might expect our government agencies to gather thorough data on how this technology can affect our health and our environment. Instead, these agencies have refused to gather crucial safety information, and have worked to limit, not promote, our access to information about what we are eating.

From Massey, R., "The ABC's of GE Food: The Science and Regulations," TomPaine.com. With permission.

http://www.tompaine.com/opinion/2001/04/18/1.html

THE LABELING LOGIC

No Labels, No Tests, No Problem Says FDA

Charles Margulis is a Genetic Engineering Specialist with Greenpeace.

In January 2001, the Food and Drug Administration (FDA) issued two new proposals for genetically engineered food. Neither proposal makes any significant change from the agency's 1992 position. In 1992, FDA's policy allowed unlabeled, untested gene-altered foods into our diets and environment for the first time. Even before then, scientists, doctors and the American people criticized FDA for allowing this mass food experiment.

The first of the new FDA proposals is an agency rule that still fails to require any pre-market safety testing of genetically engineered (GE) food. The rule only requires that companies notify the agency before bringing a new engineered food to market. Instead of mandatory testing and review, FDA will allow GE foods on the market with no comprehensive scientific review.

The second proposal outlines agency guidelines for food companies who wish to "voluntarily" label their products. While agency guidelines do not have the force of law that an FDA rule would have, the net effect of the guidelines will be the same: responsible food companies who avoid GE food will have a difficult time labeling their non-GE products, while thousands of unlabeled GE foods will remain on our supermarket shelves.

Guidelines for No Labeling: Don't Ask, Can't Tell

In just the past few years, the food and biotech industry's refusal to keep GE soy, corn and other crops separate throughout the food chain has brought us thousands of foods in our supermarkets made with ingredients from these gene-altered crops. Despite massive public support for labeling, FDA has refused to require labels on any of these GE foods. Since genetic engineering first appeared in our food, any company could have "voluntarily" labeled their GE foods. Yet there is not a single GE food that has been voluntarily labeled, since food-makers are aware that consumers don't want GE food.

Now the FDA proposes guidelines for "voluntary" labeling. But if no food company has yet instituted voluntary labels on GE foods, what good are these new guidelines? FDA's guidelines are not intended to encourage food companies to label their GE foods, but to threaten responsible food companies who avoid GE food and who wish to label their non-GE products.

This is not the first time FDA has used this strategy to favor the biotech industry over consumer protection. In 1993, when Monsanto brought its genetically engineered cow drug, bovine growth hormone (BGH) to market, FDA published guidelines (written by an agency official who formerly worked as a lawyer for Monsanto) that warned milk and dairy producers that "No BGH" labels would be misleading. Several dairies and food companies, including Ben & Jerry's and Stonyfield Farms yogurt were sued for informing their consumers with "No BGH" labels..

The agency's new guidelines on GE food suggest that food producers who label non-GE products could be misleading consumers about their foods, since FDA claims that GE foods are equivalent to non-GE foods. FDA says food producers should not label products as "non-GE," since that could be misleading. But FDA does not find it misleading if GE foods are unlabeled. The agency says that there is no "material" reason that GE foods should be labeled.

Material Information for Consumers' Right to Know

FDA claims that since genetically engineered foods are equivalent to their natural counterparts, there is no "material" justification for requiring labels. A change in the nutrition or the use of one of the most common allergens by genetic engineers would trigger labeling, says the agency. To date, FDA has found no such cases, so no engineered foods are required to be labeled.

What is a "material" change to FDA? Any "performance changes" in food, such as changes in the foods' physical properties, flavor characteristics, functional properties or shelf life have been categorized as "material" information that must be communicated to consumers via labeling. Almost all of the engineered crops in the U.S. have been altered for insect resistance (plant foods that contain a built-in pesticide that remains in the food after harvest) or for tolerance to toxic herbicides. In both cases, the GE foods are altered for specific functional changes, yet FDA refuses to acknowledge that these are material changes about which consumers have a right to know.

The agency has stated that selling irradiated food without labels would mislead consumers, since that would falsely represent to consumers that the food was not so processed. To be consistent, one would expect that FDA would also rule that unlabeled GE foods would mislead consumers, since this would lead them to believe that the food was not so altered.

Yet FDA holds a completely opposite view on genetically engineered food. In an Alice-in-Wonderland twist, the agency maintains that no labels are needed on foods made with gene-altered ingredients, but natural foods labeled as "not genetically engineered" could be misleading. In FDA's Wonderland, a consumer who buys an unlabeled tomato has a right to expect that the tomato is not irradiated, but no right to expect that the tomato is not genetically engineered.

Also, when it required labels on irradiated food, FDA noted that labeling could be required even if there were no significant changes in the food. FDA said that the fact that the food had been irradiated could be "material information" to consumers who desire such information. Clearly the vast majority of Americans desire labeling of genetically engineered food. Even a biotech industry poll found that 93 percent of Americans want labels on engineered food, and FDA's own focus groups found that "Virtually all participants said that bioengineered foods should be labeled as such. ..."

No Safety Testing: Don't Look, Don't See

To defend its failure to require pre-market testing, FDA claims that none of the gene-altered foods currently on the market are different than their natural counterparts in any nutritional or other meaningful way. But FDA refuses to acknowledge scientific differences between genetic engineering and traditional food production, and so by failing to look for changes, it falsely claims there are none.

FDA's failure to require safety testing of genetically engineered food is premised on two fundamental elements that contradict science and/or federal law. They are:

- That genetically engineered foods are "generally recognized as safe" (GRAS); and
- That genetically engineered foods are no different from foods produced through traditional breeding.

Generally Recognized as Safe

In 1958, the Federal Food, Drug and Cosmetic Act was amended to prevent the use of untested and potentially unsafe food additives. The law noted that food producers must conduct safety testing of such additives because otherwise such food companies could

> . . . endanger the health of millions by using an untested additive for as long a time as it may take for the government to suspect the deleteriousness of this additive, schedule research into its properties and effects, and, finally — perhaps years later — to begin the years-long experiments needed to prove the particular additive safe or unsafe.

Clearly the intent of the law is to require the developer of a new food ingredient to provide evidence of the safety of that food before it is allowed on the market. Yet FDA has allowed genetically engineered ingredients into the diet of millions of consumers without testing, despite the fact that these foods contain genetic material never before seen in the human diet.

An exception to the rule of food additive testing is for substances that are "generally recognized as safe." GRAS substances are defined as "... having been adequately shown through scientific procedures to be safe under the conditions of its intended use." GRAS can also apply to substances used safely in foods prior to 1958. Clearly, genetically engineered foods were not in use prior to 1958, so FDA must be relying on "scientific procedures" demonstrating their safety if it believes GE foods are GRAS.

However, the opposite is true. FDA's own scientists have warned that there is not enough scientific evidence to conclude that genetically engineered foods are GRAS. A general consensus among scientific experts is supposed to be the basis for finding a substance to be GRAS. But even FDA's lead Biotechnology Coordinator acknowledged in 1991 that "... there are a number of specific issues addressed in the [FDA policy] document for which a scientific consensus does not exist currently." He went on to note that allergies in gene-altered foods and the need for toxicity tests were among the issues of scientific concern.

Similarly, other scientists have warned of the lack of safety data on genetically engineered foods. Writing in *Science* magazine, one researcher noted that a survey of all available peer-reviewed literature on the subject found just seven studies on the health effects of gene-altered foods. A report from an expert panel commissioned by the Royal Society, Canada's most eminent scientific body, found recently that there are no "... validated study protocols currently available to assess the safety of GM [genetically modified] foods in their entirety (as opposed to food constituents) in a biologically and statistically meaningful manner."

Genetic Engineering Versus Traditional Breeding

Another pillar of FDA's policy is the assumption that the process of genetic engineering is merely an extension of traditional breeding, and that gene-altered foods have been shown to be "substantially equivalent" to their natural counterparts.

But again, FDA's policy contradicts the findings of its own scientists. Writing a summary of the agency's experts' opinions, a senior FDA scientist found: "The [draft policy] document is trying to force an ultimate conclusion that there is no difference between foods modified by genetic engineering and foods modified by traditional breeding practices ... [but] the processes of genetic engineering and traditional breeding are different, and according to the technical experts in the agency, they lead to different risks." Another FDA scientist wrote: "We should also keep in mind that plant genetic engineering is an entirely new adventure with potentially new effects."

Because genetic engineering uses genes that breeders never have used, in ways that breeders never have used, to create organisms that breeders could never create, FDA scientists found that the potential of the technology "... is beyond the realm of possibility of standard breeding practice. The food safety of organisms derived from recombinant DNA technologies do not have the history of the safe use that has come to be associated with organisms derived by standard breeding practices." While FDA continues to falsely assert that the genes used in GE crops are well-characterized and common in food, in fact the technology allows the use of novel genetic material in cross-species combinations to produce foods that have never before been part of our diet or our environment.

Still, FDA says that biotech companies have shown that their products are "substantially equivalent " to natural foods. Since gene-altered foods are equivalent, the agency says, it would be an unfair burden to require more testing or labeling than would be required for the traditional counterpart.

But there is no scientifically agreed upon procedure to measure "substantial equivalence." FDA found that Monsanto's gene-altered soybeans were equivalent to natural soy. Yet tests showed the genetically engineered beans had significantly lower levels of protein and one fatty acid, significantly lower levels of phenylalanine, an essential amino acid, and 27 percent more trypsin inhibitor, a potential allergen that interferes with protein digestion and has been associated with enlarged cells in rat pancreases. Since "substantial equivalence" is not a scientific concept, FDA overlooks these and other statistically significant differences between natural and gene-altered foods.

---

http://www.tompaine.com/news/2001/04/24/index.html

Thumbs Down on FDA Rules for Biotech Food — Proposed Regulations Unnecessary, Harmful

From: Competitive Enterprise Institute
Wednesday, January 17, 2001

WASHINGTON, D.C. —

The Competitive Enterprise Institute issued a strong condemnation of proposed new regulations for genetically-engineered foods released today by the Food and Drug Administration. "This scientifically groundless proposal is an obvious capitulation to anti-biotechnology agitators," said Gregory Conko, CEI's Director of Food Safety Policy. "Biotech foods aren't inherently less safe than other foods, so singling them out for greater oversight simply can't be justified on any grounds other than interest group politics."

Dozens of scientific organizations, including the U.S. National Academy of Sciences, the Royal Society of London, the UN's Food and Agriculture Organization, and the World Health Organization have found that biotech foods do not inherently pose different risks from non-biotech plants—a point that the FDA acknowledges in its proposal. FDA also expressed confidence that its current policy "adequately address[es] both the scientific and regulatory issues" raised by biotech foods.

Nevertheless, industry advocates of agricultural biotechnology support the FDA's proposal because they believe it does not represent a fundamental change from current regulatory practice. "But this support may backfire on the industry," said Conko. "FDA's willingness to bow to pressure groups could make biotech foods an even bigger lightening rod for activists."

Under the proposed rule, manufacturers wishing to sell genetically-engineered crop plants are required to submit a pre-market notification to the FDA four months prior to marketing their products. The FDA can then extend this review by an additional four months, and it can ultimately refuse permission for the products to be marketed. "The rule grants FDA too much discretion to decide how much or how little information is necessary to satisfy critics," said Conko. "It also gives them the authority to arbitrarily withhold approvals indefinitely, regardless of safety."

Many in the biotechnology industry also believe that the public will be reassured by FDA's increased oversight. "But most of the people who are skeptical of biotech foods, simply don't realize how much testing is already conducted or how stringent the current regulations are. So making those rules tougher isn't likely to help," said Conko. "Expecting people who are uninformed today to be fully informed tomorrow is ludicrous."

Testimony of
Mary C. Sophos
Senior Vice President and Chief Government Affairs Officer
Grocery Manufacturers of America, Inc.
Comments on USDA Proposed Rulemaking

April 17, 2001

GMA Comments on USDA/GIPSA Advanced Notice of Proposed Rulemaking

Richard Hardy
Grain Inspection, Packers & Stockyard Administration
U.S. Dept. of Agriculture
1400 Independence Avenue, SW - Rm 0757-S
Washington, DC 20250-3650

Re: Docket: FGIS-2000-001a
(RIN 0580-AA73)
Comments in Response to USDA/GIPSA ANPRM

Mr. Hardy:

The Grocery Manufacturers of America (GMA) appreciates the opportunity to provide the U.S. Dept. of Agriculture/Grain Inspection, Packers & Stockyards Administration (GIPSA) with the following response to its Advance Notice of Proposed Rulemaking (ANPRM) (FR 65; 71272-71273; 11/30/00).

The future of the U.S. grain handling system lies clearly in shipping an increasing volume of U.S. commodities to growing economies abroad, but also in supplying a diverse array of value-added agri-food products to food manufacturers and ultimately consumers domestically. GMA applauds GIPSA's recognition of these emerging changes in the commodity marketplace and its desire to anticipate and better serve the U.S. agri-food community.

Grocery Manufacturers of America
As background, GMA is the world's largest association of food, beverage and consumer product companies. With U.S. sales of more than $460 billion, GMA members employ more than 2.5 million workers in all 50 states. The organization applies legal, scientific and political expertise from its member companies to vital food, nutrition and public policy issues affecting

the industry. Led by a board of 44 Chief Executive Officers, GMA speaks for food and consumer product manufacturers at the state, federal and international levels on legislative and regulatory issues. The association also leads efforts to increase productivity, efficiency and growth in the food, beverage and consumer products industry.

The majority of GMA members do not market the commodity products mentioned in the ANPRM. Most have quality control and assurance programs, however, that involve various levels of product tracking, segregation and testing to ensure the integrity of the incoming supplies of such products for processing into well-known, brand-name products. These specifications can change daily, if not hourly and the marketplace has continually risen to the challenge of meeting these customer-supplier demands, with relatively few accommodations including those for new technologies such as biotechnology.

Nevertheless, GMA members support the potential for biotechnology to improve the quality, safety and attributes of their products. Within the decade, GMA members envision the introduction of new value-added commodities developed through biotechnology that will provide processing efficiencies or benefit consumers directly. The ongoing segmentation and changes in commodity markets will accelerate due, in large part, to the introduction of these value-added biotech products. Whether or not such commodities are biotech commodities, however, is just one parameter that companies presently look at when determining product specifications and cannot be singled out or looked at in isolation.

General Comments

In response to GIPSA's ANPRM, GMA believes that:

- The well-developed body of U.S. legal principles on food safety, food labeling and food advertising should apply to all agri-food products. In contemplating future actions, GIPSA should focus its efforts equitably on ensuring the integrity of marketing and handling systems for all agri-food products, and not just those developed through biotechnology.
- At the most basic level, validated test methods should accompany U.S. Government review of individual biotech products to assist in the identification and marketing of such products.
- Sufficient expertise and resources exist in the private sector to respond to additional marketplace demands (e.g., new testing methodologies) in a timely manner, as the marketplace has demonstrated historically.
- It would be counterproductive to adopt specific rules about identity preservation (IP) systems and certification that would lock current practices and currently-available methods in place. Technological developments would, in all probability, quickly render them obsolete. This is particularly true if no consideration is given to the development of value-added products that one day will benefit consumers in the differentiated products of GMA member companies.

- Market mechanisms, such as price incentives, will go far to ensure that the quality and added value of agri-food products with consumer benefits are captured within IP systems. Standards of performance will be product-specific, between customer and supplier, and not based on broad commodity availability and standards, which currently justify a U.S. Government role.

Specific Questions
In response to specific questions raised by the agency, GMA provides the following comments:

As more certifying companies and organizations evolve to review and verify the performance of food company IP systems, should USDA have a role in the accreditation of these certifying companies and organizations?

As previously mentioned, GMA members see no sufficient need nor reason for a distinct set of regulations specifying the procedures for complying with laws of general application that are well-developed and which food companies have long been obliged to comply. Moreover, GMA members believe governments should not delegate law enforcement to private organizations paid by parties who have the obligation to comply.

USDA is in the process of developing a program for accrediting qualified commercial and public laboratories for the analytical detection of grains and oilseeds derived from biotechnology. Should USDA expand this program for other commercialized crops? Should USDA include laboratories outside the U.S. in its program? Would this help facilitate the marketing of U.S. crops?

The body of principles governing laboratory practices is well established, and reliable accrediting organizations already exist. Again, we see no sufficient need nor justification for an entirely distinct set of regulations and programs for agri-food products derived from biotechnology.

Should USDA provide direct analytical detection services and certification for crops derived from biotechnology? Should such involvement be limited to US crops or expanded to include imported crops?

USDA should not distinguish products of biotechnology by providing direct analytical detection services and certification for them, whether domestic or imported.

GMA appreciates the opportunity to provide this input to the Agency. Should you have any questions, please contact Karil L. Kochenderfer, Director of Environment & New Technologies with GMA at 202/295-3927.

Sincerely,

Mary C. Sophos
Senior Vice President and Chief Government Affairs Officer

Staff Contacts

- Karil L. Kochenderfer

Press Contacts

- Peter Cleary

Related GMA Documents dealing with - BIOTECHNOLOGY

BUZZ

- October 12, 2000 Americans Say Hunger More Urgent World Problem Than Global Warming and Pollution; More than Two-Thirds Support the Use of Food and Agricultural Biotechnology as a Tool to Help Solve Problem
- September 11, 2000 "The Promise of Biotech Remains Alive", by C. Manly Molpus, GMA President and CEO
- July 12, 2000 Leading Science Authorities Hail The Promise of Biotechnology
- January 11, 2000 Biotechnology -- FDA Commissioner Speaks Out on Biotech Benefits, Policy
- December 8, 1999 Leading Scientists Support Food Biotechnology

COMMENT

- May 4, 2000 GMA Petition, Food and Drug Administration, Biotech-Related Claims
- May 4, 2000 Petition For Industry Guides Concerning Claims About Modern Biotechnology In The Production Of Foods
- January 14, 2000 Biotechnology in the Year 2000 and Beyond
- May 30, 1999 Comments to FDA on Codex Alimentarius Biotechnology Task Force
- March 30, 1999 Statement by GMA on the proposed ad hoc Intergovernmental Task Force on Biotechnology

CORRESPONDENCE

- June 12, 2000 GMA Comments, USDA National Organic Food Program
- April 26, 2000 Letter to FDA Office of Nutritional Products, Labeling & Dietary Supplements, Re: Codex Committee on Food Labeling, Biotechnology Labeling

- March 21, 2000 GMA Letter to U.S. Codex Office, Re: Biotech Labeling Provisions
- February 10, 2000 Groups Urge Senators To Oppose Mandatory Biotech Labeling Bill
- January 27, 2000 35 Major Trade Groups Urge House Members To Oppose Kucinich Biotech Labeling Bill
- November 17, 1999 GMA, FMI Letter to FDA: Make Biotech Consultation Process Mandatory; Groups Reiterate Support for Agency's Biotech Policy
- November 12, 1999 In Letter to Clinton, 38 Food, Farm Organizations Urge President to Support FDA Food Biotechnology Policy

NEWS RELEASE

- May 3, 2001 FDA Should Require Testing Methods for New Biotech Crops, Says GMA
- January 17, 2001 FDA PROPOSAL FOR BIOTECH LABELING AND PRODUCT APPROVAL "A VICTORY FOR CONSUMERS"
- October 12, 2000 GMA SURVEY SHOWS AMERICANS LEARNING MORE ABOUT BIOTECHNOLOGY; FOOD CONSUMPTION PATTERNS UNCHANGED
- October 4, 2000 GMA APPLAUDS JUDGE'S REJECTION OF ACTIVISTS' LAWSUIT AGAINST FDA BIOTECH POLICY
- September 22, 2000 GMA Statement On Kraft Foods, Inc. Voluntary Taco Shell Recall
- July 19, 2000 GMA REMINDS CONSUMERS THAT BIOTECH FOODS ARE SAFE FEDERAL AGENCIES, SCIENTIFIC BODIES ATTEST TO SAFETY
- June 14, 2000 GMA: USDA Organic Seal Must Not Mislead Consumers
- May 5, 2000 GMA, Industry Groups Asks FDA, FTC To Explain "Rules Of The Road" For Non-Biotech Food Claims
- May 3, 2000 GMA: FDA's Reaffirmation Of Biotech Safety Will Help Ensure "Continued Consumer Confidence" In Regulation
- April 26, 2000 Mandatory Biotech Labels Falsely Imply Safety Concerns, Raise Trade Barriers
- April 13, 2000 GMA: Congressional Biotech Report Should Aid Effort to Dispel Fear, Misinformation
- April 5, 2000 GMA: NAS Report Reinforces The Safety Of And The Science Behind Food Biotechnology
- March 28, 2000 State Level Biotech Laws Would Lead To "Confusing Patchwork" Of Regulation, Says GMA

- March 1, 2000 Regulators Should Develop Criteria For Voluntary, Non-Biotech Food Claims; GMA's Katic Addresses OECD Biotech Conference
- February 24, 2000 GMA Expert To Speak At Major International Food Biotech Forum; OECD Conference To Address Science, Safety Issues
- February 22, 2000 Senate Biotech Labeling Bill "Would Result In Consumer Confusion, Not Education"; 35 Major Trade Groups Urge Hill To Oppose Measure
- January 31, 2000 Mandatory Biotech Labeling Bill Would Increase Food Costs, Lower Farm Prices; 35 Major Trade Groups Urge Hill To Oppose Measure
- January 21, 2000 GMA Biotech Task Force Chairman To Serve On USDA Advisory Panel; General Mills' Sullivan Part Of Panel To Advise Glickman On Biotech Policy
- January 18, 2000 Public Would Benefit From "Reader-Friendly" Info on Biotech Safety, GMA Tells FDA
- December 13, 1999 GMA: As Hearings Conclude, FDA Biotech Policy Should Remain Science-Based, Reassure Consumers on Safety
- November 18, 1999 GMA: FDA Biotech Labeling Policy Protects Consumers, "Serves The Public Interest"
- November 15, 1999 Three Dozen Food, Farm Groups Urge President to Support FDA Biotech Policy
- November 10, 1999 Mandatory Labeling Bill Would Mislead Consumers About Safety Of Biotechnology
- October 7, 1999 Groups Call Attack on Safety Review of Biotech Foods "Beyond Reason"
- October 7, 1999 New Web Site Launched on Benefits of Food Biotechnology, GMA Tells Senate Panel Alliance for Better Foods Unveils www.betterfoods.org
- May 14, 1999 GMA: PROCESSED FOODS NEED TO BE ON EQUAL FOOTING WITH AGRICULTURE AT WTO TALKS
- February 1, 1999 BRAND MARKETERS MUST PREPARE CONSUMERS FOR THE "NEXT WAVE" OF BIO-ENHANCED FOODS, SAYS GMA
- December 17, 1998 BIOSAFETY PROTOCOL COULD CHOKE OFF GLOBAL TRADE, GMA COALITION TELLS PRESIDENT Associations Urge Administration to Promote Bio-Enhanced Products
- November 5, 1998 GMA FORMS STRATEGIC ALLIANCE WITH GEORGETOWN UNIVERSITY CENTER FOR FOOD AND NUTRITION POLICY
- November 4, 1998 BIOSAFETY PROTOCOL CREATES GLOBAL BARRIERS TO TRADE, GMA TELLS WHITE HOUSE Enhanced Foods Threatened by United Nations Proposal

- July 28, 1998 FROM BIOTECH TO NUTRITIONAL CLAIMS, U.S. FOOD INDUSTRY CALLS FOR ELIMINATING RED TAPE IN EUROPEAN MARKETPLACE
- May 27, 1998 DON'T TURN BACK THE CLOCK ON ENHANCED FOODS, GMA SAYS FDA Policy on Genetically-Altered Food Based on Science, Not Fear
- March 4, 1998 GMA CALLS FOR STRONGER ALLIANCE BETWEEN FARMERS AND FOOD MANUFACTURERS

TESTIMONY

- April 23, 2001 GMA Testifies In Opposition to Massachusetts Biotech Crop Moratorium Bill
- April 17, 2001 GMA Comments on USDA/GIPSA Advanced Notice of Proposed Rulemaking
- April 11, 2001 GMA Testifies in Support of FDA-Based Maine "Biotech-Free" Labeling Bill
- March 29, 2001 Expert Witness Testifies on Behalf of GMA in Opposition to Maine Mandatory Biotech Food Labeling Bill
- March 29, 2001 GMA Letter in Opposition to Maine Mandatory Biotech Food Labeling Bill
- March 21, 2001 GMA Testifies in Opposition to New Hampshire Mandatory Food Biotechnology Labeling Bill
- March 21, 2001 GMA Testifies in Support of FDA GMO-Free Labeling Guidance at Minnesota Hearing
- February 9, 2001 GMA Letter of Opposition to Colorado Biotech Food Labeling Bill
- February 9, 2001 GMA Letter of Opposition to Rhode Island rBST Labeling Bill
- February 7, 2001 Biotech Labeling Legislation Pending In Mexico
- November 28, 2000 Perspective on Risk Assessment Of Cry9C Protein in StarLink Corn and on Adverse Event Reporting
- October 24, 2000 GMA testimony of Lisa Katic before the New York Genetically Modified Crop Working Group
- April 26, 2000 GMA Testimony, California Senate Committee on Finance, Investment, and International Trade: Implications of Biotechnology on International Trade
- April 12, 2000 Written Comments In Support of Hawaii Biotechnology Resolution, HCR 37
- March 28, 2000 California Senate Agriculture Committee, Informational Hearing on Biotechnology

- March 1, 2000 GMA Opposes Rhode Island rBST Labeling Bill, H.B. 7511
- February 29, 2000 Organisation For Economic Cooperation And Development, Edinburgh Conference On The Scientific And Health Aspects Of Genetically Modified Foods; Panel On Regulatory Frameworks And Consumer Involvement
- November 18, 1999 GMA Testimony on Biotechnology: FDA Biotech Labeling Policy Protects Consumers, "Serves The Public Interest"
- October 7, 1999 Food Industry Stands Behind Food Biotechnology As Alliance Launches New Web Site
- April 12, 1999 GMA Opposes Legislative Docket 713
- March 10, 1999 GMA Opposes House Bill 887
- July 28, 1998 The Transatlantic Economic Partnership

Related GMA Documents dealing with - LABELING

COMMENT

- March 28, 2000 Guidance on Significant Scientific Agreement

CORRESPONDENCE

- November 30, 2000 GMA Supports Canadian Move to Harmonize Nutritional Labeling
- November 12, 1999 In Letter to Clinton, 38 Food, Farm Organizations Urge President to Support FDA Food Biotechnology Policy

NEWS RELEASE

- May 4, 2001 U.S. Industry and Government Position Prevails on Biotech Labeling at Codex
- May 3, 2001 FDA POLICY KEEPS IMPORTANT HEALTH INFORMATION FROM PUBLIC, GMA TELLS CONGRESS
- January 17, 2001 FDA PROPOSAL FOR BIOTECH LABELING AND PRODUCT APPROVAL "A VICTORY FOR CONSUMERS"
- November 18, 1999 GMA: FDA Biotech Labeling Policy Protects Consumers, "Serves The Public Interest"
- November 15, 1999 Three Dozen Food, Farm Groups Urge President to Support FDA Biotech Policy
- June 11, 1999 GMA: NATIONAL UNIFORMITY FOR FOOD ACT REPRESENTS "COMMON SENSE CONSUMER INITIATIVE" House Bill Introduced by Rep. Burr; Companion to Sen. Roberts' Measure

- May 27, 1999 GMA: COMMON SENSE BILL PROVIDES UNIFORM, SCIENCE-BASED FOOD LABELING STANDARDS FOR CONSUMERS
- May 11, 1999 GMA: "TIME IS RIPE" FOR FDA TO RECONSIDER AND REVISE HEALTH CLAIMS APPROVAL PROCESS
- September 1, 1998 FOOD INDUSTRY URGES FDA NOMINEE TO PUT FOOD ISSUES ON EQUAL LEVEL

TESTIMONY

- May 7, 2001 GMA Testifies in Support of FDA Biotech-Free Labeling Guidance at New York Hearing
- May 3, 2001 FDA POLICY KEEPS IMPORTANT HEALTH INFORMATION FROM PUBLIC
- April 17, 2001 GMA Comments on USDA/GIPSA Advanced Notice of Proposed Rulemaking
- April 13, 2001 GMA Letter of Opposition to Mass. Green Dot Labeling Legislation
- April 11, 2001 GMA Testifies in Support of FDA-Based Maine "Biotech-Free" Labeling Bill
- March 29, 2001 Expert Witness Testifies on Behalf of GMA in Opposition to Maine Mandatory Biotech Food Labeling Bill
- March 29, 2001 GMA Letter in Opposition to Maine Mandatory Biotech Food Labeling Bill
- March 21, 2001 GMA Testifies in Opposition to New Hampshire Mandatory Food Biotechnology Labeling Bill
- March 21, 2001 GMA Testifies in Support of FDA GMO-Free Labeling Guidance at Minnesota Hearing
- March 21, 2001 Voluntary Labeling Indicating Whether Foods Have or Have Not Been Developed Using Bioengineering
- March 15, 2001 Letter in Opposition to Connecticut Prop 65 Clone Bill
- February 15, 2001 GMA Letter in Opposition of Nebraska Product Origin Labeling Bill
- February 14, 2001 GMA Letter in Opposition to Florida "Sell By" Product Dating Bill
- February 9, 2001 GMA Letter of Opposition to Colorado Biotech Food Labeling Bill
- February 9, 2001 GMA Letter of Opposition to Rhode Island rBST Labeling Bill
- February 7, 2001 Biotech Labeling Legislation Pending In Mexico
- February 29, 2000 GMA Testimony, Connecticut Warning Label Legislation, SB 433

- November 18, 1999 GMA Testimony on Biotechnology: FDA Biotech Labeling Policy Protects Consumers, "Serves The Public Interest"
- June 23, 1999 GMA Urges Veto Of Senate Bill 945
- May 10, 1999 GMA Opposes HB 3136
- April 12, 1999 GMA Opposes Legislative Docket 1985
- April 12, 1999 GMA Opposes Legislative Docket 713
- March 10, 1999 GMA Opposes House Bill 887
- July 28, 1998 The Transatlantic Economic Partnership

---

http://www.gmabrands.com/news/docs/Testimony.cfm?docid=741&

AGRICULTURAL BIOTECHNOLOGY – DOMESTIC REGULATION

Issue:

The Farm Bureau supports the continued development of agricultural products enhanced through biotechnology. We support increased efforts through biotechnology to increase the marketability of agricultural products, address environmental concerns, to increase farm income by decreasing input costs and improving product quality.

Background:

Food biotechnology is the use of genetic science to create new products from plants and animals. Biotechnology allows scientists to identify an individual gene that governs a desired trait, copy that gene and insert the copy into other plant cells. This genetic modification can be used, for example, to create crops that grow faster, ripen more slowly, or are resistant to disease and pests. Biotechnology is monitored by several federal agencies, including the Food and Drug Administration (FDA), USDA and EPA. The FDA considers most crops derived from biotechnology as inherently safe to eat as long as the genes are already present in the food supply.

Domestic Regulation

FDA labeling policy requires mandatory labeling if products of biotechnology differ from traditional products in terms of their composition, their nutritional content, or their allergenicity. In January, the FDA issued official guidelines regarding GM or GM-free for voluntary labeling of foods containing references to biotechnology-enhanced ingredients. FDA also, proposed a rule that would require companies to notify the FDA 120 days in advance of marketing and submit data showing the product to be safe before selling any new products of biotechnology. Farm Bureau supports the actions by FDA.

The Environmental Protection Agency (EPA) has regulatory authority for insect-protected crops because they produce their own protection against pests (i.e. Bacillus thuringiensis, or Bt crops). EPA regulates environmental exposure to these crops to ensure there are no adverse effects to the environment or any beneficial, nontargeted insects and other organisms. EPA is currently revising its policy, known as the Plant-Incorporated Protectant (PIP) rule, on how to regulate these types of crops. All Bt crops will undergo a reregistration safety review this year.

The Department of Agriculture (USDA) oversees field-testing of biotech crops. USDA has recently undertaken the certification of testing and testing methods for biotech crops.

The administration has recently established an interagency task force to coordinate positions and activities of all federal agencies related to biotechnology.

AFBF Policy:

Farm Bureau supports agricultural biotechnology. US government agencies (FDA, USDA and EPA) should continue to serve their roles in providing unbiased, scientifically based evaluations concerning the human and animal safety and wholesomeness, as well as the environmental impacts of biotechnology-enhanced products.

We urge state and national political leaders to develop a positive national strategy for biotechnology research, development and consumer education. Part of this strategy should include an open and frank dialogue with all interested parties. We believe that our competitive advantage in world markets will be maintained only by the continued support and encouragement of echnological advancements. U.S. government agencies, particularly the USDA and the Food and Drug Administration (FDA), should continue to serve their respective roles in providing unbiased, scientifically-based evaluations concerning the human and animal safety and wholesomeness, as well as the environmental impacts of biotechnology-enhanced commodities.

We favor strong patent support to encourage these new technologies. Patents should be broad enough to provide reasonable protection of development costs, but should not be so broad as to grant one developer the right to a whole class of future developments. We support seed tags on packages of agricultural seed stock that warrant genetic purity of seed contained therein. We will also support legislation which allows producers to recover all damages in those instances where the seed does not conform to the genetic purity indicated on the seed tag. Adequate and accurate information on acceptable markets, and market and planting restrictions must be provided in writing to producers prior to the time they purchase the original input product.

Agricultural products that are produced using approved biotechnology should not be required to designate individual inputs or specific technologies on the product label.

We support:

1. The science-based labeling policies of FDA, including:
   (a) no special labeling requirement unless a food is significantly different from its traditional counterpart, or where a specific constituent is altered (e.g., nutritionally or when affecting allergenicity); and (b) voluntary labeling using statements that are truthful and not misleading; and
2. Voluntary labeling of identity-preserved agricultural and food products that is based on a clear and factual certification process.

Action:

Legislation is expected in the 107th Congress. Farm Bureau supports the actions taken by the FDA and will likely oppose legislation introduced to require mandatory labeling or attempts to treat foods produced through biotechnology as food additives.

The Farm Bureau has encouraged the administration to takes steps to coordinated government efforts in support of biotechnology and insure continued consumer confidence and marketability of biotech products domestically and abroad. Farm Bureau has recommended that the administration designate a White House lead to coordinate all administration messages and efforts (across all agencies) on biotechnology.

May 2001

Section III

Media Documents

The first example of the journalistic treatment of the genetic engineering of foods, from *Newsweek*, typifies how the subject has been presented in the popular print media. One can almost tick off systematically the criteria for newsworthiness discussed in Chapter 4 of the text. From its title, "Frankenstein Foods?" to its concluding paragraph anticipating "heat" at the upcoming WTO meeting in Seattle, the piece is a paradigm of the genre.

After the intimations that something monstrous may be involved are suggested by its title, the article begins, as so many news stories do, with an anecdote about an obscure Parisian intellectual who is challenging a transnational corporation, McDonald's, and, by extension, the forces of globalization. Within just a few paragraphs, the reader is presented with an international conflict — "a hybrid of cultural and agricultural fears" — that threatens American agricultural exports. That globalization itself is undergoing something of a trial, and that the "little guy" seems to be winning, only makes the "story" more engaging.

In presenting this conflict, the piece also satisfies other criteria of newsworthiness. The situation is, of course, novel, since the international reaction against genetically altered foods is of recent vintage even though Americans have been eating them for almost a decade. That it deals with food also makes it "newsworthy": nothing could be more "proximate" or more intimately relevant. What we have, then, is a classic story of a new, potentially insidious threat to our food supply that is already exposing cultural fault lines between Europe, especially England, and the U.S., and foreshadowing some problems with the world's plunge into globalization. What could be better? Thus, though there has been no evidence of death or illness from the consumption of a genetically modified food, drama and emotion are nevertheless evoked. And both sides are dutifully represented. Rebecca Goldburg from Environmental Defense and Gordon Conway from the Rockefeller Foundation are each called upon to talk about the "science," nascent as it is with respect to this subject. Even celebrities find their way into the story, as Prince Charles and Paul McCartney involved themselves in the controversy.

The two other media pieces reproduced here are aimed at smaller, more partisan audiences. Maria Margaronis' "The Politics of Food" from the liberal-oriented *The Nation* is almost exclusively concerned with the clash of cultures underpinning the issue, and plays up Monsanto's failure to market GMOs in England despite its massive public relations effort and its subsequent concessions to European fears in order to maintain profit — something of a David and Goliath story. On the other hand, "Food Risks and Labeling Controversies," by Henry Miller and Peter VanDoren, in *Regulation*, almost completely focuses on the "science" of the issue in an attempt to calm fears and anxieties by making clear not only the absence of any cases of human death or harm but even of any allegations of potential risks to human health. It highlights what it regards as the irrationality of such fears by pointing to the wide public consumption of herbal remedies that have been subject to less testing and more intimations of harm. While the broad purpose of the article in *The Nation* is to stir people up, or at least engage them, the goal of the story in *Regulation*, published by the Cato Institute, a right-wing think tank, is to calm people down, hence the more cerebral treatment.

Discussion

It might be profitable to compare *Newsweek's* treatment of the subject with that of other general circulation magazines. *Time* published an article on it in its June 19, 2000 issue called "Will Frankenfood Feed the World?" and *U.S. News and World Report* took on the subject in an article called "The Curse of Frankenfood" in its July 26, 1999 issue. Apparently the lure of headlining articles on genetically modified foods with the nickname that Europe has pinned on them — Frankenfood — is too tempting to resist. Even the more sober *Business Week* called its piece "Furor Over 'Frankenfood.'" What is the *business* point here? The *environmental* point? Is false fear being fostered by suggesting that genetically modified food is science gone mad, even in the face of no concrete evidence? What are the consequences for environmental policymaking?

Note that in virtually all these articles, and in a follow-up story in *U.S. News and World Report*, "Engineering the Harvest," which appeared in its March 13, 2000 issue, the focus is on the European/American clash, and the subtheme is the possible loss of necessary nutrition to third world inhabitants due to the unavailability of genetically modified foods. What does bringing in "starving" nations add to the issue? Why is it such a prominent part of so many stories?

Finally, it will no doubt be enlightening to follow to its logical consequence the point so prominent in the Cato article. What does it suggest that herbal compounds are flying off supermarket and "health food store" shelves, despite their not having tested safe. They do, of course, sport a label that tells the consumer as much. Are labels, even those that tell you that what you are about to consume has not survived systematic laboratory testing, consoling?

Some other journal articles worth looking at for comparisons are: "Seeds of Change," in *Consumer Reports*, September 1999, "Brave New Food," in the April/May 2000 issue of *Mother Earth News*, and "The Great Yellow Hype," in *The New York Times Magazine*, March 4, 2001.

Frankenstein Foods?

That's what Europeans are calling genetically modified crops that abound in America. Exporters have been forced to listen.

By Kenneth Klee

Don't look for the southern French town of Montredon on your globe. It isn't even on local road maps, perhaps because it has only 20 inhabitants. But one of them, a Parisian intellectual turned activist-farmer named José Bové, may change that. He's the leader of the mobs of farmers who've trashed several McDonald's lately. Last week, with 200 supporters chanting outside the jail, Bové declined a Montpellier court's offer of bail and remained behind bars, the better to spotlight his cause. And that would be? "To fight against globalization and advance the right of people to eat as they see fit," he explained. Grievance No. 1: the U.S. desire to export genetically modified crops and foods.

So far, so French, right? But spin that same globe to Peoria, Ill., home of U.S. agribusiness giant Archer Daniels Midland. There, even as Bové's judges readied their decision, the self-styled "supermarket to the world" was demonstrating that the customer is, indeed, always right. In a fax to grain elevators throughout the Midwest, ADM told its suppliers that they should start segregating their genetically modified crops from conventional ones, because that's what foreign buyers want. It didn't matter that GM crops are widely grown by U.S. farmers, and that there's no evidence that the taco chips and soda you're enjoying right now are anything worse than fattening. ADM had noticed something new sprouting under the bright, warm sun of economic interdependence: a strange hybrid of cultural and economic fears. So it decided to act before the problem got any bigger.

Public opposition to GM foods in Europe has been mounting for more than two years, especially in Britain and France. Both Prince Charles and Paul McCartney have come out against the stuff. Now the protests and the tabloid headlines about "Frankenstein Foods" have reached such a pitch that they're reverberating across the Atlantic. Secretary of Agriculture Dan Glickman, a longtime backer of biotechnology, admitted as much in a key speech in July. So did Heinz and Gerber when they announced the same month that they'll go to the considerable trouble of making their baby foods free of genetically modified organisms. Groups such as Greenpeace, which have long fought biotech on both continents, are crowing. U.S. trade officials, who face a tough fight keeping markets open for American agricultural products, are worrying. And U.S. consumers, who have never really thought much about genetically modified foods, are just plain confused.

As well they might be, given the vastly different experiences the United States and Europe have had. In the United States, the FDA issued a key ruling in 1992 that brought foods containing GM ingredients to market quickly, and without labels. Companies such as Monsanto introduced herbicide-resistant soybeans and corn that makes its own insecticide. U.S. farmers loved the products; by 1988, 40 percent of America's corn crop and 45 percent of its soybeans were genetically modified. In Europe, meanwhile, there was no real central regulator to green-light the technology and allay public concerns, and many more small farmers for whom biotech represented not an opportunity but a threat. Leaders have tried to steer a course between encouraging a new industry and giving the voters what they want, including labeling rules.

So, to each his own, right? Not in 1999. If Europe is selling America Chanel perfume and Land Rovers, America will want to sell Europe its soybeans and corn—and maybe even its fervent faith in progress. While European biotech companies such as Novartis avoided the limelight, St. Louis-based Monsanto decided to press its case. The timing was terrible. GM fears were already running high last summer when Monsanto ran an informational campaign; Britain's 1996 bout with mad-cow disease, though unrelated, had weakened European confidence in regulators and industrial-strength agriculture. Monsanto's PR effort only made the mood worse, as have a string of bad-news food headlines since then: dioxin-contaminated chicken in Belgium last spring; tainted Coke in Belgium and France this summer, and a punitive U.S. tariff on imports of foie gras and other products, imposed in July because Europe won't accept American hormone-fed beef.

That last, also nongenetic, dispute actually triggered the vandalism at McDonald's last month. But to many of France's famously irascible small farmers, it's all of a piece. Even among the broader public in France and Britain, the GM-foods issue seems to be intersecting with second thoughts about globalization. French farmer protest American imperialism. But just last week their biggest customers, grocery giants Carrefour and Promodes, announced a $16.5 billion merger that will position them well in a global battle with America's Wal-Mart—and put further cost pressures on farmers. Britain is a hotbed for Internet start-ups. But Brits still tune in to the BBC radio soap "The Archers" to see if young Tommy will go to jail for helping a group of eco-warriors wreck a GM-crop trial site on his uncle's land.

Would an American jury let Tommy go? Probably not. Consumers Union, whose Consumer Reports magazine features a big piece on GM foods this month, has put together an array of poll data suggesting Americans would like to see GM food labeled but remain interested in its benefits. Of course, if Tommy's trial were held in Berkeley, Calif., where the school board has announced a ban on GM foods, he might walk.

U.S. activists, encouraged by the successes of their European brethren, hope to build on such sentiments. Some of the rhetoric is extreme, and one group—or perhaps it's just one person—has resorted to vandalism, trashing a test-bed of GM corn at the University of Maine last month and crediting the act "Seeds of Resistance." But there's science going on, too. A Cornell University study published in the journal Nature in May found that half of a group of monarch-butterfly caterpillars that ate the pollen of insecticide-producing Bt corn died after four days. What if the pollen spreads to the milkweed the monarchs lay their eggs in? "The arguments aren't enough to say we shouldn't have any biotechnology," says Rebecca Goldburg of the Environmental Defense Fund. "But they are enough to say we should be looking before we leap."

Of course we should, says Gordon Conway, president of the Rockefeller Foundation and an agricultural ecologist. Invited to speak to the Monsanto board in June, he used the forum to talk about the need to go a little slower. But, he adds, don't worry about the monarch. Bioengineers can stop the pesticide (which is *supposed* to kill caterpillars; they eat the corn) from being expressed in pollen. "There are always problems in the first generation of a new technology," he says. And, he adds, successes. The foundation just unveiled a genetically modified rice grain it funded to improve nutrition in the developing world. If a shouting match over GM foods should derail such not-for-profit efforts, he says, "that would be a tragedy."

Agriculture Secretary Glickman doesn't see Americans growing as fearful as Europeans, mainly because he thinks Americans have more faith in their regulators. He also thinks that labeling of GM foods is a big part of the answer—not mandatory labeling, which industry opposes and activists demand, but voluntary labeling. "I'm not going to mandate this from national government level," he told Newsweek, "but I believe that more and more companies are going to find that some sort of labeling is in their own best interest." Especially companies that want to export.

Because, as ADM showed with its heartland-stopping announcement on Thursday, it isn't only up to Americans anymore. Brian Kemp, a Sibley, Iowa farmer, made an urgent call to his elevator on Thursday to see if it would still buy his GM corn. It will—this year. "Europe is so important to the industry that it could mean we'll really have to pull back on growing GM crops in this country," says Walt Fehr, head of Iowa State University's biotech department. "Given the choice, who wants to grow GM?"

Glickman says the trade issue—which is sure to generate plenty of heat at the November World Trade Organization meeting in Seattle—will be a tough one to resolve. "But I think over the next five years or so we can get it done." That's a mighty slow pace, considering how quickly the industry came along in the previous half decade. But then, you generally do travel faster when you travel alone.

September 13, 1999 Newsweek 33-35

With John Barry in Washington, Scott Johnson in Montpellier, Jay Wagner in Des Moines, William Underhill in London and Elizabeth Angell in New York

Maria Margaronis, "The Politics of Food"

The Nation, December 27, 1999

> Case sawed shakily at his steak, reducing it to uneaten bite-sized fragments, which he pushed around in the rich sauce. "Jesus," Molly said, her own plate empty, "gimme that. You know what this costs?" She took his plate. "They gotta raise a whole animal for years and then they kill it. This isn't vat stuff."
>
> **—William Gibson, Neuromancer**

LONDON

A year ago, Monsanto chairman Robert Shapiro had the future in his pocket. His vast "life sciences" corporation was at the cutting edge of the new agricultural revolution, genetic modification; the spread of GM seeds throughout the United States, he told his shareholders, was the most "successful launch of any technology ever, including the plow." The little manner of European distaste for the new crops would, he felt sure, be resolved by the right kind of PR and some careful scientific reassurance. As Ann Foster, the company's personable British flack, patiently explained to anti-GM campaigners here, "people will have Roundup Ready soya, whether they like it or not."

So far, things have not gone according to plan. The European Union has a de facto moratorium on the commercial growing of GM crops, pending further discussion (the only exception is the Swiss company Novartis's Bt com, currently being grown in Spain). Austria, Luxembourg, Italy and Greece have total or partial bans on the technology. Even the Blair government, in love with the sleek promises of high-tech business and keen to keep Clinton sweet, has bowed to public pressure and put off the commercial planting of GM seeds in Britain for at least three years. (Environment Minister Michael Meacher, whose views on the subject are carefully tracked by the CIA, has reportedly said in private that GM crops will never be grown commercially here.) Shoppers have rejected GM food in droves, prompting a breathless race among the supermarket chains to go GM-free. As a report by the Britain government's Science and Technology Committee put it, "At the current rate at which food manufacturers are withdrawing GM ingredients from their products, there will be no market for GM food in this country."

US soy exports to Europe are down from $2.1 billion in 1996 to $1.1 billion in 1999, and anxiety about GM crops (or genetically engineered crops, as they're generally known in the United States) is blowing across the prairies. Last spring and summer a series of reports by the influential Deutsche Bank urged investors to pull out of agricultural biotechnology altogether: "The term GMO [genetically modified organism] has become a liability. We predict that GMOs, once perceived as the driver of the bull case for this sector, will now be perceived as a pariah." In October a chastened Shapiro apologized to Greenpeace for his "enthusiasm," which, he acknowledged, could be read as "condescension or indeed arrogance." Monsanto's stock has gone seriously pear-shaped, and the board has reportedly considered a company breakup.

What happened? How did a loose assemblage of European environmental activists, development charities, food retailers and supermarket shoppers stop a huge multinational industry, temporarily at least, in its tracks?

The first protests against genetic modification took place in America in the late seventies, when activists from a group called Science for the People destroyed frost-resistant strawberries and delayed the construction of Princeton's molecular-biology building. Then they fizzled out. Americans, by and large, trust the FDA to keep the levels of toxicity in their daily bread down to a psychologically manageable level and don't worry too much about the source of the goodies that fill their horn of plenty. The great grain factories of the Midwest work their magic far from the places most people visit to enjoy nature. In much of Europe, though, nature and agriculture go hand in glove, occupying the same physical and social space. Europe's layered patchwork of farming and culinary landscapes has taken shape over 2,500 years, altered by small and large migrations, the conquest and loss of colonies, wars and revolutions. Europeans feel strongly about what they eat: Food is a matter of identity as well as economy, culture as well as nurture.

The most dramatic changes in European farming in this century came about partly as a result of the experience of famine during World War II: The much-reveled Common Agricultural Policy (CAP) of the European Union has its origins in the determination that Europe should never again see mass starvation. By protecting and supporting their farmers against the vagaries of trade while simultaneously investing in intensive agriculture (a contradiction in terms, you might say, since roughly 80 percent of Europe's farm subsidies go to 20 percent of its farmers), European governments hoped to insure long-term food security for their people. But, as they usually do, the contradictions eventually came home to roost.

"The fourth agricultural revolution," says Tim Lang, professor of food policy at Thames Valley University and one of the new food movement's intellectual lights, "is beginning just as the third one—agrochemicals and intensive farming—is unraveling." The unraveling has made itself felt both in the economic crisis that affects many of Europe's farmers and in a series of food-safety scandals caused by deregulation and overintensive production. The outbreak of bovine spongiform encephalopathy (BSE) in Britain's cattle in the eighties and its appearance in humans as the fatal new-variant Creutzfeldt-Jakob disease in the nineties was the most powerful catalyst for the public's loss of faith in governments and food producers. In one terrifying package, BSE tied together the new "economical" farming practices (in this case the feeding of ground-up cow carcasses to cattle), the easing of health and safety standards, and government's willingness to lie for the food industry even at the cost of human lives.

So far, new-variant CJD has killed forty-three people in Britain; the chief medical officer recently warned that millions may still contract it from beef they ate fifteen years ago. By some estimates, the whole affair has cost about $6.5 billion, much of it put up by the European Union. Elsewhere in Europe, similar stories break with depressing regularity. Last summer, for instance, a cover-up of dioxin contamination in animal feed brought down the Belgian government and part of the Dutch Cabinet and had worried gourmets across the continent throwing out chickens, eggs and Belgian chocolate to the tune of $800 million. (The Coca-Cola crisis that followed, in which 30 million cans and bottles of the elixir of life were poured down the drain after a number of people reportedly fell ill, turned out to be a genuine case of mass hysteria.) The anxiety is only partly contained by sideshows like the Anglo-French beef war, in which the British agriculture minister decided to boycott French food in retaliation for France's refusal to lift its ban on British beef with the rest of the European Union—simultaneously publicizing an EU report that found sewage sludge processed into French animal feed. The happy tabloid trumpeting that ensued momentarily restored the beef of Old England to its rightful place as a bulwark against the filthy Frogs, allowing the Daily Mail to boost its circulation with pictures of cows in berets and toilet-paper necklaces amid cries of "Just say Non!"

The biotech companies danced into this minefield with all the grace of an elephant in jackboots.

Ten years ago, agricultural biotechnology was debated only by what Labor MP Joan Ruddock (former leader of the Campaign for Nuclear Disarmament) calls "men in white coats and men in gray suits," with environmental NGOs like Greenpeace and Friends of the Earth reporting on their activities but mounting no large-scale protests. In 1990 the first GM additive approved for use in British food, a GM baker's yeast, was swallowed without qualms; so was the GM tomato paste sold by Sainsbury's supermarket in 1996, at a lower price than its conventional equivalent. The trouble started that same year when the American Soybean Association, Monsanto and the US trade associations told British food retailers that they could not—would not—segregate American GM soybeans from the conventional kind, undermining the golden rule of consumer-friendly capitalism: Let them have choice. Around the same time, media and public awareness of the issue reached critical mass, and the supermarkets started getting worried letters from their customers asking them not to use GM ingredients. The arrogance with which the American biotech firms approached the European food industry is the stuff of legend. Bill Wadsworth, technical manager of the frozen-food chain Iceland, recalls a meeting in September 1997 at which a biotech executive actually said, "You are a backward European who doesn't like change. You should just accept this is right for your customers." A few weeks later Wadsworth was on a plane to Brazil, where he found a grower and processor of non-GM soybeans and began to set up a vertically integrated supply chain for Iceland's processed foods. Iceland began to raise the issue's profile with its customers, pointing out that while Iceland's foods were GM free, those of other supermarkets were contaminated. Before long every supermarket chain in the country was inundated with mail and phone calls about GM food and had begun to follow suit. In June 1998 a poll showed that 95 percent of British shoppers thought that all food containing GM ingredients should be labeled.

Meanwhile, the field testing of GM crops in Britain by Monsanto, AgrEvo, Novartis and other companies gave a dramatic focus to the environmental arguments against genetic modification. Media-savvy eco-activists in decontamination suits or grim reaper outfits began to pull up trial plantings and leaflet supermarkets; by the summer of 1998, hardly a week went by without reports of some new, inventive, nonviolent protest. English Nature, the government's own environmental watchdog, and the Royal Society for the Protection of Birds both added their authoritative voices to calls for a moratorium on planting, citing the unpredictable and uncontainable dangers of releasing the new organisms into the ecosystem. Gene transfers could produce herbicide-resistant "superweeds"; crops genetically engineered to be toxic to insects might well affect the whole food chain, further damaging the biodiversity of a landscape already impoverished by intensive farming. In a country where the membership of environmental and conservation groups outstrips the membership of political parties by four to one, the disappearance of cornflowers and skylarks from fields and hedgerows is a political issue. Prince Charles's entry into the fray on the side of the green campaigners did much to enhance the post-Diana credibility of a man who not so long ago was widely ridiculed for talking to his plants.

By the time Monsanto launched its too-clever-by-half ad campaign to sell biotechnology to the British public in the summer of 1998, the bonfire had been prepared. The united front of environmentalists, shoppers and food retailers, animated in part by fury at the hubris of multinationals trying to pull the wool over their eyes, was joined by an army of development NGOs outraged by Monsanto's efforts to corner Third World seed markets with a technology that could destroy farmers' livelihoods while pretending to "feed the world."

The spark that lit the flames was the broadcast that August of a television documentary about the work of Dr. Arpad Pusztai, a researcher at a government-funded institute who claimed that feeding GM potatoes to laboratory rats had slowed their growth and damaged their immune systems. Dr. Pusztai rapidly lost his job amid assertions that his work was flawed and incomplete, but the whole affair catapulted GMOs into the tabloid firmament. With its usual brash enthusiasm The Express launched a populist crusade against "Frankenfoods," and pretty soon not a man, woman or child in Britain was left in the dark. The GM controversy even made The Archers, BBC radio's venerable daily soap about an English farming family: To the relief of fans everywhere, young Tommy Archer was recently found not guilty of criminal damage after destroying a test crop of GM oilseed rape in one of his uncle's fields.

Downing Street has remained largely unmoved by all this protest, allowing Tory leader William Hague (who has himself been caricatured as a genetically modified vegetable) to make political hay out of Labor's urban unconcern for the environment and dazzled obeisance to the biotech firms. To Tony Blair, pro-business to his toenails, the GM revolution is part of the white heat of new technology that will carry the British economy through the next century. In the words of the government's Chief Scientific Adviser, Sir Robert May, "We have played a hugely disproportionate part in creating the underlying science: are we going to lose it like we lost things in the past?" Dolly the sheep, after all, was cloned here.

If we do "lose it" in the long run, it will be in part because of the government's serious misreading of the public mood. Had they proceeded from the start in an

"Your people have rejected GM food," said Vivek Cariappa, an organic farmer from southern India who is active in his country's thriving anti-GM movement. "Where will it go? It won't go into the sea. It will go to countries like ours." With careful honesty, Ruddock explained to the farmers that their British colleagues, on the whole, don't share their concerns: "Britain has been run as multinational farming enterprises with subsidies from the CAP. It is mostly people in urban areas, pressure groups, pushing for change in agricultural practice, except for a small organic minority." When Juli Cariappa asked if Britain really wants to leave its food basket in the hands of the multinationals, Ruddock paused, looked in her eye, and said, reluctantly, "Yes."

If the biotech companies have their way we could soon be on course for William Gibson's nightmare future, in which the rich eat real food grown by artisan farmers and the poor eat genetically engineered "vat stuff" when they eat at all. As long as food is treated as a commodity like any other and traded to maximize profits, there is little chance of a reduction of world hunger or of a significantly safer diet for the fortunate few. As Tim Lang puts it, "We have to see that it is the production of food that matters, not just its consumption." Or, in the crisp words of José Bové, "We are faced with a real choice for society. Either we accept intensive production and the huge reduction in the number of farmers in the sole interests of the World Market, or we create a farmer's agriculture for the benefit of everyone." The shape-shifting global coalition that tripped the advance of genetically modified crops in Europe and staged the carnival of protest in Seattle has its work cut out for it. But the genie is out of the bottle. Food—which in its progress from seed to stomach links ecology, labor, poverty, trade, culture and health—will be a key item on the menu of the next century's struggles for democracy against the arbitrary power of the giant corporations.

Maria Margaronis is a Nation contributing editor living in London. Thanks to D. D. Guttenplan for additional reporting on this piece.

Reprinted with permission from the December 27, 1999 issue of *The Nation.*

John Stauber, "Food Fight Comes to America"

The Nation, December 27, 1999

As the international uprising against genetically engineered (GE) foods continues to grow, the worst fear of US government and business officials is that the commotion abroad will awaken Americans, who unknowingly already consume biotech foods being rejected in Europe. The victories of their foreign counterparts, meanwhile, are providing fresh inspiration for US food activists, some of whom have struggled for decades to win media coverage, citizen attention and regulatory action. The Food and Drug Administration has officially opposed biotech food labeling and mandatory safety testing since 1992 [see Kristi Coale, facing page]. But now that Europeans are forcing American companies to segregate and label genetically engineered foods, it is much more difficult to claim that the same can't be done in the United States.

Last summer was a watershed event for many US farmers, who planted Monsanto's biotech corn and soybeans, only to find them rejected abroad. Some are shifting back to traditional varieties, at least until the crisis is resolved. Gary Goldberg, CEO of the American Corn Growers Association, suggested in November that farmers avoid genetically engineered seed corn and try to obtain non-engineered varieties before farmer demand depletes supplies of old-fashioned seed.

The US food and biotechnology industries are now in full "crisis management" mode, their PR experts and lobbyists working furiously to prevent the same kind of defeat suffered on foreign shores. One example is the recently launched Alliance for Better Foods, run from the DC office of the PR/lobby firm BSMG, which also represents Monsanto and Phillip Morris, America's largest food company. Monsanto's PR firm Burson-Marsteller recently bused 100 members of a Washington, DC, Baptist church to stage a pro-GE-foods rally outside an FDA hearing. But if events in Europe are any guide, the momentum may have shifted to a new alliance of grassroots environmentalists, consumer activists and family farmers. The Los Angeles Times noted in October that "a storm of protest has reached US shores, leading some experts to predict that agricultural biotechnology could go the way of nuclear energy—falling out of favor because of public fears and unfavorable economics."

The key to any successful biotech "issue management" campaign is repeating simple but carefully chosen messages that can set the terms of the debate. This was true with Monsanto's genetically engineered bovine growth hormone (rBGH), administered to cows to increase milk output. In the case of rBGH, one message was that "the milk is the same." This isn't true, and changes in milk are a reason the drug hasn't been approved by Europe or Canada. But the message worked here, where, after a furious PR and lobbying campaign, the FDA approved the use of rBGH and allowed sales of dairy products without consumer labeling. Six years later, Monsanto claims that one-third of US cows are in herds injected daily with rBGH.

Another simple but effective PR tactic, known as "the third-party technique," puts messages in the mouths of independent-seeming experts, such as scientists and doctors, whom journalists and the public are more likely to trust. Besides government "watchdogs" at the FDA and the US Department of Agriculture (USDA), such messengers can include former Surgeon General C. Everett Koop, the AMA and its prestigious publication Journal of the American Medical Association, and the American Dietetic Association. All these and more have vouched for rBGH, and we can expect an avalanche of similar trusted experts reassuring us about biotech foods in the months ahead. Meanwhile, many right-wing pro-industry groups have launched their own PR campaigns against the "fearmongering" of consumer and environmental activists. At Thanksgiving, for example, the National Center for Public Policy Research faxed to newsrooms a release headlined Activists Attack Bio-Engineered Food Despite Benefits to the Poor and the Sick. All these tactics would fail, of course, if the media did their job by thoroughly investigating and reporting the issue of genetically engineered foods, and that is why media management is the number-one goal of every PR campaign. As its ultimate weapon, industry has successfully lobbied into law "agricultural product disparagement" statutes that give them new powers to sue people who criticize their products. The first such lawsuit was filed in Texas against Oprah Winfrey and her guest Howard Lyman for the crime of airing a public debate on mad cow disease and its risks in the United States. A jury ruled in Oprah's favor, prompting her to crow that "free speech rocks." The reality is that her case is on appeal, and she has spent more than $2 million thus far in legal bills that she will never get back. Food-disparagement statutes survive intact in Texas and twelve other states, and this shot across the bow of the media has already had a chilling impact on coverage of other food controversies.

One of the smartest moves by Monsanto in the rBGH fight was hiring Carol Tucker Foreman, an influential and well-connected Democratic insider and lobbyist. Previously, Foreman had been the executive director of the DC-based Consumer Federation of America and then, under President Jimmy Carter, an Assistant Secretary of Agriculture. Soon after her stint at USDA she launched her own DC lobby firm with many corporate clients. While paid by Monsanto to lobby for rBGH, Foreman also coordinated the Safe Food Coalition, whose members include a number of big Washington-based nonprofits such as Consumer Federation of America and the Center for Science in the Public Interest. Running the coalition allowed Foreman to maintain dual identities as both a consumer advocate and a corporate food lobbyist. Earlier this year Foreman left her lobby firm and returned to the Consumer Federation of America, where she now says she favors labeling genetically engineered foods.

Another major Washington food lobbyist is Michael Jacobson, the executive director of the Center for Science in the Public Interest, an advocacy group dubbed "the food police" by industry for its attention-getting news conferences against unhealthy fats and sugars in the American diet. When it comes to biotech foods, however, CSPI has been less vigilant, failing to oppose Monsanto's rBGH during the long struggle over its approval. Jacobson now frets that mandatory labeling of genetically engineered food sold in supermarkets, as called for in a bill introduced in November by Representative Dennis Kucinich, could kill a goose he hopes will lay genetically engineered golden eggs such as "increased yields, reduced toxins, increased nutrient levels, and modified fatty acid composition."

If inside-the-Beltway groups like CSPI and CFA are conflicted and unlikely to lead the charge to gain mandatory safety testing and consumer labeling of GE foods, who is? A broad array of seasoned activists has been fighting this battle for a long time, among them author Jeremy Rifkin, attorneys with the Center for Food Safety, the Rural Advancement Foundation International-USA, the National Family Farm Coalition, the Council for Responsible Genetics, Consumers Union (publisher of Consumer Reports) and the Union of Concerned Scientists. In meetings this year a number of these organizations and others formed the Genetic Engineering Action Network, which is united around four objectives: mandatory safety testing of GE foods, mandatory consumer labeling if they pass safety tests, long-term industry liability to cover unforeseen problems and an end to the domination of food and agriculture by "supermarket to the world" companies.

No one involved in the US fight expects it to be quick or easy. Says Ronnie Cummins of the Organic Consumers Association, "We have to do what Europe and Japan have done—build a powerful organized movement of farmers, consumers and environmental activists who will target and boycott companies." Recently the FDA held three public hearings on genetically engineered foods. Critics call them staged events and dog-and-pony shows, but they have provided a media forum for advocates of safety testing and labeling. The activists want biotech foods off the market entirely until a rigorous system of health and ecological testing has been devised. Like their colleagues in Europe, they are promoting the Precautionary Principle—the common-sense maxim of "looking before you leap"—as the basis for public policy. Adherence to the Precautionary Principle would obviously have dire consequences for companies whose bottom-line profits depend on selling as many biotech foods as quickly as possible, but it seems a minimal level of protection against the inevitable unforeseen consequences of genetically engineering the world's food supply.

John Stauber is executive director for Media & Democracy and founder of PR Watch, a quarterly journal that investigates corporate and government propaganda (www.prwatch.org). He is co-author of Toxic Sludge Is Good For You! and Mad Cow USA, both published by Common Courage.

Market-based alternatives to more government regulation of foods and dietary supplements

Food Risks and Labeling Controversies

BY HENRY I. MILLER

AND PETER VANDOREN

Questions about food safety and regulation abound, among them: How safe are biotech foods—foods derived from gene-spliced organisms? Should they be labeled? Should herbal dietary supplements continue to be exempt from federal regulation on safety, effectiveness, and labeling?

The European Union passed legislation in 1998 requiring labeling to identify all foods containing genetically modified (i.e., gene-spliced) ingredients, which caused large retailers to remove all such foods from their shelves. Responding to those events and to intimidation by Greenpeace, two the United States' largest producers of baby food, Heinz and Gerber, have announced that they will use only non-biotech ingredients in their products. Demonstrators in Europe and the United States have protested the marketing of biotech foods. And fearing that many or most U.S. consumers will reject biotech foods, some U.S. farmers have canceled orders for genetically engineered seeds.

The professional risk analysis community believes that biotech foods are just more precisely constructed versions of plants engineered with other long-established techniques. Mandatory labeling of foods to indicate the presence of gene-spliced products would incorrectly signal to consumers that the government believes there is something to worry about—or, at least, that there is something fundamentally different about such products. The Food and Drug Administration's oversight of biotech foods—which is based on potential risk, not the use of certain techniques—is appropriate and adequate to ensure food safety.

In contrast, the risk-analysis community is alarmed by the state of virtual anarchy in the market for herbal supplements. Many of the products are known to be toxic, carcinogenic, or otherwise dangerous (ephedra and chaparral, for example), although only a few supplements, including saw palmetto for treating enlarged prostate glands and gingko biloba for enhancing memory in Alzheimer's patients, have been shown to be efficacious. There is no shortage of information available to consumers about dietary supplements, but it is heavy on advocacy and light on scientific proof.

Nevertheless, the lack of scientific evidence for dietary supplements' safety and effectiveness seems not to faze many consumers, who spend a fortune on unproven nostrums and jeopardize their health using dangerous ones. Known, serious side effects include blood-clotting abnormalities, high blood pressure, life-threatening allergic reactions, cardiac arrhythmia, exacerbation of autoimmune diseases like arthritis and lupus, and kidney and liver failure. Even persons who believe that the process for approval of new drugs is too stringent argue for controls on dietary supplements.

In view of the recent—and continuing—attacks on biotech foods, we begin by assessing the lack of scientific merit in those attacks. We then turn to the question of labeling for biotech foods and dietary supplements. We argue that market forces can serve consumers' interests, obviating the need for additional Food and Drug Adminstration (FDA) scrutiny of biotech foods or dietary supplements.

The Scientific Consensus In 1986, the Paris-based Organization for Economic Cooperation and Development (OECD) issued *Recombinant DNA Safety Considerations*, in which OECD's Group of National Experts on Safety in Biotechnology found that

> genetic changes from gene-splicing techniques will often have inherently greater predictability compared to traditional techniques, because of the greater precision that the gene-splicing technique affords; [and] it is expected that any risks associated with application of gene-spliced organisms may be assessed in generally the same way as those associated with non-gene-spliced organisms. (p. 31)

A landmark 1989 report of the U.S. National Research Council, *Field Testing Genetically Modified Organisms: Framework for Decisions*, went even further, observing that "with organisms modified by molecular methods, we are in a better, if not perfect, position to predict the phenotypic expression" (p. 13). That statement expresses the scientific consensus that our ability to predict "phenotypic expression"—the very essence of risk assessment related to environmental protection and public health—is superior for gene-spliced foods.

In 1993, OECD's Group of National Experts specifically addressed food safety, concluding in *Safety Evaluation of Foods Derived by Modern Biotechnology* that

> evaluation of foods and food components obtained from organisms developed by the application of the newer techniques does not necessitate a fundamental change in established principles, nor does it require a different standard of safety. (p. 13)

In the same report, the group of experts described the concept of "substantial equivalence" in new foods. The concept—a form of regulatory shorthand—applies to those foods that do not raise safety issues that require special, intensive, case-by-case scrutiny. (The U.S. delegation suggested the use of "substantial equivalence," which is borrowed from FDA's definition of a class of new medical devices that do not differ materially from their predecessors and, thus, do not raise significant regulatory concerns.)

OECD has continued to explore the concept of substantial equivalence. In 1998, another expert group concluded in *Report of the OECD Workshop on the Toxicological and Nutritional Testing of Novel Foods* that

> while establishment of substantial equivalence is not a safety evaluation *per se*, when substantial equivalence is established between a new food and the conventional comparator [antecedent], it establishes the safety of the new food relative to an existing food and no further safety consideration is needed. (p. 15)

Fallacies and Conspiracy Theories Some recent attacks on biotech foods have been based on a misinterpretation of a laboratory experiment involving the monarch butterfly and on flawed experiments that purportedly showed toxicity in rats fed gene-spliced, lectin-enhanced potatoes. But a more fundamental attack is one on substantial equivalence by Erik Millstone, Eric Brunner, and Sue Mayer in their article, "Beyond 'substantial equivalence'," which appeared last year in *Nature*.

Millstone et al. call substantial equivalence a "pseudo-scientific concept because it is a commercial and political judgment masquerading as if it were scientific" (p. 526). Wholly ignoring empirical experience and scientific consensus, Millstone et al. suggest that gene-spliced foods should be treated "in the same way as novel chemical compounds, such as pharmaceuticals, pesticides and food additives, and [requiring] a range of toxicological tests, the evidence from which could be used to set acceptable daily intakes (ADIs)" (p. 526). Then, of course, we would need "regulations … to ensure that ADIs are never, or rarely, exceeded" (p. 526).

By considering all changes arising from gene splicing—but only those changes—as novel, Millstone et al. ignore the fact that many products on the market are derived from "wide crosses"—hybridizations in which genes are moved from one species or one genus to another to create a variety of plant that does not and cannot exist in nature. They demand extensive, difficult-to-perform, hugely expensive testing of foods from gene-spliced plants, but not of other foods from the dozens of new plant varieties produced by traditional techniques of genetic modification, such as hybridization, that enter the marketplace each year without premarket review or special labeling.

If new and draconian regulatory regimens are necessary for the new biotechnology, they are certainly applicable to traditional biotechnology as well. And in that regard, one must wonder how we would calculate ADI for the mutant peach called a nectarine or the tangerine-grapefruit hybrid called a tangelo. Such an exercise would clearly be absurd. And where it is not absurd—as when estimating the acceptable intake of foods such as potatoes and squash known to have high endogenous levels of natural toxins—the exercise has nothing to do with the method of genetic manipulation used to construct the plant.

In sum, the argument advanced by Millstone et al. illustrates the fallacy that underlies many of the unscientific attacks on the new biotechnology—the assumption that somehow gene splicing introduces into organisms (and the foods derived from them) greater uncertainty or risk than older, less-precise genetic-modification techniques. Yet, neither scientific consensus nor empirical evidence supports that view. As *Nature* editorialized in 1992,

> the same physical and biological laws govern the response of organisms modified by modern molecular and cellular methods and those produced by classical methods. ... [Therefore] no conceptual distinction exists between genetic modification of plants and microorganisms by classical methods or by molecular techniques that modify DNA and transfer genes. (Vol. 356, p. 1)

FDA Policy Although FDA does not use the term "substantial equivalence" in food regulation, it applies the concept in its risk-based policy toward "new plant varieties."

FDA does not routinely subject foods from new plant varieties to premarket review or to extensive scientific safety tests. Instead, it considers that the usual safety and quality-control practices used by plant breeders—mostly chemical and visual analyses and taste testing—are generally adequate for ensuring food safety.

FDA's policy defines certain safety-related characteristics of new foods that, if present, require special scrutiny by the agency. Those characteristics include the presence of a substance that is completely new to the food supply, an allergen presented in an unusual or unexpected way (for example, a peanut protein transferred to a potato), a change in the level of a major dietary nutrient, or an increase in the level of a toxin normally found in food.

A product's composition, characteristics, or history of use may suggest the need for additional testing. For example, potatoes usually are tested for the glycoalkaloid solanine toxin because it has been detected at harmful levels in some new potato varieties that were developed with conventional genetic techniques.

The absence of characteristics correlated with heightened risk, in effect, defines a product that is substantially equivalent to antecedent products. FDA does not subject such food to premarket review, whether the plant arose by gene splicing or "conventional" genetic-engineering methods.

A SCIENTIFICALLY CONSISTENT RISK POLICY WOULD MAINTAIN THE CURRENT FDA policy toward biotech foods but subject the supplement industry to additional scrutiny. But rather than argue that policy should be governed by scientific understanding—as opposed to unsound hysteria propagated by interest groups and political actors—we argue that market forces may obviate the need for more government intervention.

We focus, therefore, on whether markets can supply information that enables consumers to make informed choices about the risks they face in biotech foods and dietary supplements. Can we identify the conditions under which markets will supply information about products as part of the normal competitive process? Can we identify the conditions under which Gerber and Heinz, for example, will choose to market two or more types of food: one advertised as non-gene-spliced and higher priced, the other unpromoted, unlabeled as to its biotech ingredients, and sold at a lower price? Also, can "certified" dietary supplements command higher prices than "uncertified" supplements? If the answers to these questions are "yes," markets can serve everyone's preferences, no matter how misguided those preferences may be.

A Primer on Product Differentiation Firms will provide detailed information about their products in an attempt to distinguish them from other firms' products so long as it is profitable to do so—that is, so long as the extra price they can charge for the information exceeds the cost of providing it. A market condition known as *separating equilibrium* occurs if firms can offer different products with different amounts of information at different prices simultaneously. A *pooling equilibrium* occurs if firms cannot sustain markets for differentiated products at different prices.

To illustrate a separating equilibrium, let us assume that products are not labeled: consumers cannot determine whether dietary supplements are safe or effective or what they contain, nor can they determine the presence or absence of gene-spliced ingredients. Suppose one firm can then make more money by offering labeled food or dietary supplements. Will other firms be forced to follow, or will the market support a variety of products and information, including products with no information? If the market can support different products at different prices with differing levels of information, a separating equilibrium is possible.

A pooling equilibrium can be one of two types. In one instance, firms will not label their products if consumers are not willing to pay enough to cover the extra costs of providing information (including extra production, handling, and packaging costs). Alternatively, firms may find it unprofitable to market unlabeled products because most consumers fear the worst and are willing to pay more for labeling.

The decisions by Gerber and Heinz to use non-gene-spliced ingredients suggest that both firms believe that a biotech-free pooling equilibrium is inevitable in the market for baby food; that is, firms will not profitably be able to offer both biotech-free and unlabeled baby food. The experience of a British firm supports that view. J. Sainsbury PLC, a British supermarket chain, in 1996 began selling a bioengineered tomato puree that was labeled voluntarily. Initially, the product sold well because its price was lower than that of conventional tomato purees. But sales of the bioengineered tomato puree fell as the European public became concerned about genetic modification, and Sainsbury has withdrawn the product from the market.

An important counterexample comes from U.S. experience with recombinant bovine somatropin (rbST), a biotech version of bST—a hormone that stimulates milk production in cows. In spite of widespread concern about rbST, which led to a temporary congressional moratorium on its introduction, econometric analysis of consumer behavior after the end of the moratorium found no evidence of short-term or long-term reluctance to consume rbST milk.

Should the U.S. Government Intervene? The government should consider mandatory labeling only when a compelling case can be made that a market will fail in the absence of labeling. Suppose, for example, that every firm in a market—even those that have invested in differentiating their products—would go bankrupt if there were a scandal about contamination. Thus, in the not at all unlikely event of a scandal in the dietary supplement market, the pivotal question would be whether the conscientious firms could survive or whether the public would be unable (or unwilling) to distinguish the "good" firms from the "bad" ones?

Consider the analogous case of runs on banks. Runs occur if consumers lose faith in all banks when some banks go bankrupt. Professor Charles Calomiris has argued in *Regulation* (Vol. 22, No. 1) that consumers could differentiate "good" banks from "bad" banks even during the depths of the depression. That is, the market for information about banks worked better during the depression than conventional wisdom suggests it did. Bank information separated rather than pooled. Thus, government regulation and labeling (e.g., federal deposit insurance) arguably are not needed to avert runs on banks.

Similarly, producers of non-gene-spliced and organic foods and safe and effective dietary supplements can differentiate themselves from other producers without government mandates. Witness the growth of the Whole Foods and Wild Oats supermarket chains, both of which recently announced bans on gene-altered foods. Whole Foods, with 103 stores in 22 states and the District of Columbia, and Wild Oats, with 110 stores in 22 states and British Columbia, provide detailed information about their suppliers and products in an effort to assure consumers that they are buying genuine, high-quality, organic food products. To be sure, products sold by Whole Foods and Wild Oats cost more than similar products offered by conventional grocery stores, but Whole Foods and Wild Oats provide the type of food and information that some consumers want—and for which they are willing to pay.

IN SUM, THERE IS NOT A CLEAR CASE FOR FDA INTERVENTION IN U.S. MARKETS for food products or (perhaps) dietary supplements. In particular, FDA's present risk-based policy toward biotech foods is sound and should not be changed. But there are other motives and beliefs at work in the world, which U.S. firms must heed.

If Europeans want to consume local, more expensive, non-gene-spliced foods, who are we to say that they should consume our cheaper, more precisely crafted biotech foods? By requiring the labeling of biotech foods, European governments evidently believe there will be a pooling equilibrium in which local non-gene-spliced foods would be undifferentiated from unlabeled (and presumably imported) foods. But if European governments are accurately reflecting the sentiments of European consumers, there is likely to be a separating equilibrium in which all unlabeled foods will lose market share to foods certified as local and non-gene-spliced. Under such circumstances, even if American producers win the political fight against mandatory labeling, most unlabeled American foods would not survive in Europe.

But current European sentiments may actually be analogous to American public sentiment toward rbST, which also was very negative at first. The American market now exhibits a classic separating equilibrium: although most dairy products are unlabeled, some premium, niche products (e.g., Ben and Jerry's ice cream) are labeled as rbST-free.

Should Gerber and Heinz have yielded to anti-biotech activists? Sainsbury's experience with biotech-based tomato puree suggests that there may not be a market for labeled biotech baby food—at least in Europe. And perhaps baby food is one of the markets in which unlabeled food would not sell because consumers—who want "the best" for their babies—would suspect that there are biotech ingredients in unlabeled products and would therefore buy only biotech-free labeled products. But

Gerber and Heinz should consider the possibility that all customers, regardless of their preferences, can be served by products that differ in price and ingredients. There may be no more need for Gerber or Heinz to have a one-size-fits-all product than there is for the government to impose a one-size-fits-all regulation on food producers.

Finally, there is no pressing need for the additional government regulation of dietary supplements. Producers of supplements could contract voluntarily with a foundation that would operate like Underwriters Laboratories (UL)—a large, non-profit organization that tests and certifies products, many of which are potentially hazardous to life and property. The UL certification offers assurance of safety but not of effectiveness, except in special cases where the two are inextricably linked (e.g., fire extinguishers and smoke detectors).

The adoption of similar, third-party certification by makers of dietary supplements would protect the manufacturers' long-term interests and integrity. Most important, self-regulation would assure consumers that certified products meet certain standards of purity, potency, and quality, while preserving consumers' freedom to choose non-traditional medical therapies.

readings

- Lorna Aldrich and Noel Blisard. "Consumer Acceptance of Biotechnology: Lessons from the rbST Experience." *Current Issues in Economic of Food Markets* (Agriculture Information Bulletin No. 747-01). Washington, D.C.: U.S. Department of Agriculture, December 1998.
- John E. Losey, Linda S. Rayor, and Maureen E. Carter. "Transgenic Pollen Harms Monarch Larvae." *Nature* 399 (1999): 214.
- Debora MacKenzie. "Unpalatable Truths." *New Scientist* 162 (1999): 18.
- Ehsan Masood. "£ 150m Tax Break for Research in UK budget." *Nature* 398 (1999): 98.
- Erik Millstone, Eric Brunner, and Sue Mayer. "Beyond Substantial Equivalence." *Nature* 401 (1999): 525.
- National Research Council. *Field Testing Genetically Modified Organisms: Framework for Decisions*. Washington, D.C.: National Academy Press, 1989.
- "US Biotechnology Policy" (unsigned editorial). *Nature* 356 (1992): 1.
- Organization for Economic Cooperation and Development. *Recombinant DNA Safety Considerations*. Paris: OECD, 1986.
- Organization for Economic Cooperation and Development. *Safety Evaluation of Foods Derived by Modern Biotechnology*. Paris: OECD, 1993.
- Organization for Economic Cooperation and Development. *Report of the OECD Workshop on the Toxicological and Nutritional Testing of Novel* Foods, SG/ICGB(98) 1. Paris: OECD, 1998.
- U.S. Food and Drug Administration. "Statement of Policy: Foods Derived from New Plant Varieties." *Federal Register* 57 (1992): 22984.

Henry I. Miller is a senior research fellow at Stanford University's Hoover Institution on War, Revolution, and Peace. He was the founding director of the U.S. Food and Drug Administration's Office of Biotechnology, 1989-93, and a member of the Organization for Economic Cooperation and Development's Group of National Experts on Safety in Biotechnology, 1984-92. Peter VanDoren is the editor of *Regulation*. His publications include *Chemicals, Cancer, and Choices: Risk Reduction through Markets* (Washington, D.C.: Cato Institute, 1999) and "The Effects of Exposure to Synthetic Chemicals on Human Health: A Review," in *Risk Analysis* 16 (1996): 367.

Section IV

Science Documents

One of the more important scientific studies on the genetically modified food issue — at least politically important — was conducted by the National Academy of Sciences. The study is, of course, too long to reproduce here, and almost surely too technical for all but other scientists to understand and evaluate. But the press release summarizing its findings and the opening statement at the news conference announcing its release are reprinted here. These documents are, in most ways, more significant than the study itself, for they are what was reported in the press and, hence, what the public was led to understand as the definitive scientific truth.

In sum, the report found no evidence that transgenic plants were unsafe to eat or that they posed any particular risks to health or the environment. That was the scientific judgment. But it did recommend that the base of research upon which public policy has to be made be bolstered, and that better coordination between the regulatory agencies responsible for overseeing food safety be strengthened.

In his opening statement to the news conference at which the results were released, Perry Adkisson, Chair of the Committee conducting the study, conceded that public concern was one of the factors motivating the study, that the study group's composition was balanced, that the researchers looked more closely at potential risks than benefits, and that the product, not the process by which it was produced, was, and should remain, the focal point of study.

Organizations of scientists themselves differ on questions of safety. The American Council on Science and Health, for example, has consistently argued that honest science would allay any fears, while the Union of Concerned Scientist, an opponent of genetically modified foods, finds reason for concern in an experiment at Cornell that found potential threats to monarch butterflies from a pesticide from an engineered crop. The two articles included here represent an aspect of the scientific debate that spills over into politics.

Also included here in an effort to illustrate the problems encountered when science meets policy in an Internet debate between Dr. Peter Montague, Director of the Environmental Research Foundation, and Philip Stott, a professor of Biogeography at the University of London. It was initiated by a piece critical of genetically modified foods by Dr. Montague, which precedes the debate itself. "Faceoffs," such as the one here, are increasingly common, particularly on the Internet.

The final document in the section is a splendid paper on applying the precautionary principle to bioengineered food. A discussion of the precautionary principle as the mechanism for reconciling the conflict between scientific uncertainty and public policymaking closes the chapter on the politics of science in *Environmental Politics: Interest Groups, the Media, and the Making of Policy.*

Discussion

The National Academy of Sciences is a respected body of scientists that has no agenda of its own, thus often making it the preferred arbiter of disputes with a scientific base. But that identity itself not infrequently forces the Academy into scientific conclusions that are so carefully balanced and qualified that they fail to achieve the purpose of putting issues to rest. Studying the press release and conference at which their study of the issue was released — about what was talked about and what was not — reveals much about the role of science in public policymaking. To what extent, for example, was this public event an attempt to assuage both sides, while maintaining the standing of the Academy itself?

Even more interesting were the press and Internet accounts derived from it. The *Washington Post* of April 6, 2000, for example, headlined its article, "Biotech Crops Appear Safe, Panel Says." The headline on the Nando Media News Service that same day said, "Advantages of Biotech Food Outweigh Risks." The homepages of groups with a strong advocacy position saw other things. The April 5 news release of Environmental Defense, a major environmental group, highlighted the Academy's caveats: "Scientific Panel Calls for Stronger Controls on Biotechnology," while the biotech industry site, BIO, simply "supported" the findings of the Academy. Finally, the *Chemical and Engineering News*, trade publication of the American Chemical Society, in its April 10, 2000 issue, affirmed that "[GMO) Foods are safe, NRC report says, but stronger U.S. regulatory system is needed to avoid problems in the future." Whether these differing "spins" put on the findings reflect the sources or the judicious balancing of issues by the Academy is a key question, and one that has implications for environmental policy when science is called upon to resolve disputes.

That scientists can disagree, radically, is demonstrated by the pieces from *Nucleus,* a publication of the Union of Concerned Scientists, and the homepage of the American Council on Science and Health. That rift is even more evident in the Montague/Stott debate.

All of these documents, which represent only a tiny sample of those involved in the controversy, show how even in matters of science, to which the public turns for guidance, ambiguity and dispute remain. What does all this suggest about the possibility of basing environmental policy on "sound science?"

Does the ultimate answer lie with the "precautionary principle?"

Date: April 5, 2000
Contacts: Bill Kearney, Media Relations Associate
Megan O'Neill, Media Relations Assistant
(202) 334-2138; e-mail <news@nas.edu>

FOR IMMEDIATE RELEASE

U.S. Regulatory System Needs Adjustment
As Volume and Mix of Transgenic Plants Increase in Marketplace

WASHINGTON -- Even given the strengths of the U.S. system governing transgenic plants, regulatory agencies should do a better job of coordinating their work and expanding public access to the process as the volume and mix of these types of plants on the market increase, says a new report from the National Academies' National Research Council. The committee that wrote the report emphasized it was not aware of any evidence suggesting foods on the market today are unsafe to eat as a result of genetic modification. And it said that no strict distinction exists between the health and environmental risks posed by plants genetically engineered through modern molecular techniques and those modified by conventional breeding practices.

The committee called on the U.S. Environmental Protection Agency (EPA), U.S. Department of Agriculture (USDA), and Food and Drug Administration (FDA) to quickly come to an agreement on each agency's role in regulating plants that have been genetically modified to resist pests. It also said that any new rules should be flexible so they can easily be updated to reflect improved scientific understanding.

"Public acceptance of these foods ultimately depends on the credibility of the testing and regulatory process," said committee chair Perry Adkisson, chancellor emeritus and distinguished professor emeritus, Texas A&M University, College Station. "The federal agencies responsible for regulating transgenic plants have generally done a good job, but given the current level of public concern and following our review of the data, it is the committee's belief that the agencies must bolster the mechanisms they use to protect human health and the environment. However, I must also emphasize that we believe it is the properties of a genetically modified plant -- not the process by which it was produced -- that should be the focus of risk assessments."

As the volume of transgenic products increases, more research will be needed to examine and better detect their effects on human health and the environment so that the agencies will have a more refined scientific basis for making decisions, the committee said.

Improving Pest Resistance

Farmers have been trying to minimize their losses from crop pests for hundreds of years by using conventional breeding practices, such as hybridization, to develop crops with desirable traits. Some types of worms cause an estimated $7 billion in crop losses per year in the United States; the damage from insects is even more severe. In the past two decades, scientists have used the tools of advanced molecular biology to more precisely alter plants to be pest resistant. Scientists use these methods to introduce genes that endow plants with pesticidal traits, creating what are known as transgenic pest-protected plants. These genes may come from similar, sexually compatible species or from completely unrelated organisms. Transgenic plants have been grown commercially since 1995, and their use has increased dramatically since then. In 1999 alone, more than 70 million acres of transgenic crops were planted in the United States.

But some scientists and members of the public have expressed concern that the genetic engineering of plants could result in unsafe foods, do irreparable harm to beneficial organisms, and spur the uncontrollable growth of weeds. Given the dramatic increase in commercial planting of genetically engineered crops and the safety concerns they raise, the Research Council decided to initiate a review of the scientific data on potential health and environmental risks and the use of this data in the regulatory process.

Health-Related Concerns

Thus far, only in very rare circumstances have pest-protected plants caused obvious health or environmental problems. For example, although a human allergic reaction to a new gene product has never been documented for a commercially available transgenic pest-protected plant, one such incident did occur at the research stage. In that study, people with a known allergic reaction to Brazil nuts experienced a similar reaction when they were exposed in skin-prick tests to soybeans containing a gene transferred from the Brazil nut.

Priority should be given to developing improved methods for identifying potential allergens, specifically focusing on new tests relevant to the human immune system and on more reliable animal models, the committee said. Changes in plant physiology and biochemistry should be monitored during the development of pest-protected plants. And because the potential exists for transgenic plants to have increased levels of toxic plant compounds, EPA, USDA, and FDA should create a coordinated database that lists information about natural plant compounds of dietary or toxicological concern, to aid researchers who monitor concentrations of these compounds in such plants.

Environmental Concerns

In examining ecological concerns, the committee looked at the possibility that transgenic plants could affect organisms which are not the target of the pesticidal trait, the potential transfer of novel genes from one type of plant to another, and the evolution of new strains of immune pests.

Both conventional and transgenic pest-protected crops could impact so-called nontarget species, such as beneficial insects, but that impact is likely to be smaller than that from chemical pesticides, the committee said. In fact, when used in place of chemical pesticides, pest-protected crops could lead to greater biodiversity in some geographical areas. The committee called for more research to examine these issues.

The highly publicized report of monarch butterflies being poisoned by pollen from genetically engineered corn is an example of an issue that needs to be researched further and will require rigorous field evaluations, the committee said. In that particular report, researchers showed that pollen from corn which had been genetically engineered to produce **Bacillus thuringiensis** (Bt) toxins -- a type of insecticide -- slowed the growth, and sometimes killed, monarch caterpillars when enough pollen was placed on milkweed leaves fed to them in a laboratory. Follow-up studies are needed in the field where pollen density might be lower and the toxin might be deactivated by environmental factors.

Concern also surrounds the possibility that genes for resisting pests might be exchanged among cultivated crops and their weedy relatives, potentially exacerbating weed problems -- a high-cost nuisance for farmers and potential threat to the ecosystem. The committee recommended further research to identify plants with weedy relatives, to assess rates at which pest-resistance genes might spread, and to develop techniques that decrease this likelihood.

Another ecological concern is the potential for pests to evolve and develop a resistance to plants that have been genetically modified to kill them. The committee concluded that the ability of pests to adapt and develop resistance should continue to be evaluated. Such an occurrence could have a number of potential environmental and health consequences, including a return to the use of more harmful chemical pesticides. Strategies to manage the development of resistance in pests should be encouraged for all uses of a pesticide, be it in a spray form or produced by a plant.

Improving the Regulatory Framework

To improve coordination, EPA, USDA, and FDA should develop a memorandum of understanding for regulating transgenic pest-protected plants that identifies regulatory issues within the purview of each agency as well as issues for which more than one agency has responsibility, the committee said. The memorandum also should establish a process to ensure appropriate and timely exchange of information between agencies. For 14 years, the agencies have formulated policies for genetically modified foods under guidelines set forth in the 1986 Coordinated Framework for the Regulation of Biotechnology. The framework gives each agency a role in setting safety standards based on legal jurisdictions at the time. But the committee said the scope of each agency's oversight needs to be clarified, especially when a new product is to be reviewed by more than one agency.

Additionally, the committee took issue with exemptions in EPA's proposed 1994 rule for regulating certain transgenic pest-protected plants. EPA proposes to grant categorical exemptions for all plants that have been given a new gene from a sexually compatible plant, and for plants expressing proteins that are derived from a virus, known as viral-coat proteins. But in the first instance, the committee said that in some cases the transfer and manipulation of genes between sexually compatible plants could potentially increase human and environmental exposure to high levels of toxins. Secondly, while plants with viral-coat proteins may be safe to eat, there are environmental issues to consider because of their potential to crossbreed with weedy relatives. The committee urged EPA to reconsider its plans to grant categorical exemptions for these transgenic plants.

The committee also recommended that the agencies monitor ecological impacts of pest-protected crops on a long-term basis to ensure the detection of problems that may not have been predicted from tests conducted during the registration and approval process.

A more open and accessible regulatory process is needed to aid the public in understanding the benefits and risks associated with transgenic pest-protected plants, the committee concluded. To increase access to the process, existing Web sites for the coordinated framework should be expanded to include more detailed information and to link all of the agencies' decisions for any particular product.

The committee's work was funded by the National Research Council, which is the principal operating arm of the National Academy of Sciences and the National Academy of Engineering. The three, along with the Institute of Medicine, constitute the National Academies. They are private, nonprofit institutions that provide science, technology, and health policy advice under a congressional charter. A committee roster follows.

Read the full text of **Genetically Modified Pest-Protected Plants: Science and Regulation** for free on the Web, as well as more than 1,800 other publications from the National Academies. Printed copies are available for purchase from the **National Academy Press Web site** or at the mailing address in the letterhead; tel. (202) 334-3313 or 1-800-624-6242. Reporters may obtain a pre-publication copy from the Office of News and Public Information at the letterhead address (contacts listed above).

NATIONAL RESEARCH COUNCIL
Board on Agriculture and Natural Resources

Committee on Genetically Modified Pest-Protected Plants

Perry Adkisson* (chair)
Distinguished Professor Emeritus
Department of Entomology, and
Chancellor Emeritus
Texas A&M University
College Station

Stanley Abramson
Member
Arent Fox Kintner Plotkin & Kahn, PLLC
Washington, D.C.

Stephen Baenziger
Eugene W. Price Professor
Department of Agronomy
University of Nebraska
Lincoln

Fred Betz
Senior Scientist
Jellinek, Schwartz & Connolly Inc.
Arlington, Va.

James Carrington
Professor
Institute of Biological Chemistry
Washington State University
Pullman

Rebecca Goldburg
Senior Scientist
Environmental Defense
New York City

Fred Gould
William Neal Reynolds Professor
Department of Entomology
North Carolina State University
Raleigh

Ernest Hodgson
William Neal Reynolds Professor
Department of Toxicology
North Carolina State University
Raleigh

Tobi Jones
Special Assistant for Special Projects and Public Outreach
Department of Pesticide Regulation
California Environmental Protection Agency
Woodland

Morris Levin
Professor
Center for Public Issues in Biotechnology
University of Maryland Biotechnology Institute
Baltimore

Erik Lichtenberg
Professor
Department of Agricultural and Resource Economics
University of Maryland
College Park

Allison Snow
Associate Professor
Department of Evolution, Ecology, and Organismal Biology
Ohio State University
Columbus

RESEARCH COUNCIL STAFF

Jennifer Kuzma
Study Director

* Member, National Academy of Sciences

Genetically Modified Pest-Protected Plants: Science and Regulation

National Research Council

News Conference
April 5, 2000

Opening Statement
by
Perry Adkisson

Chancellor Emeritus and Distinguished Professor Emeritus,
Texas A&M University,
College Station, Texas
and
Chair, Committee on Genetically Modified Pest-Protected Plants

Good morning and welcome to those of you in the room, and to the reporters who are joining us by telephone. My colleagues and I are delighted to be with you today for the public release of our report, ***Genetically Modified Pest-Protected Plants: Science and Regulation.***

The National Academies have been deeply engaged in the issues surrounding genetically modified organisms from the very beginning, when these advances in molecular biology were first emerging. Those of you who have covered the biological sciences and science policy for some time may recall the 1975 conference, later to become known as Asilomar, where scientists came together for several days to discuss how they could self-regulate genetic research, once it had become understood that gene splicing, indeed, was feasible. The conference was organized by the National Academy of Sciences, and it laid out the guiding principles for using recombinant DNA techniques in the laboratory.

A little over a decade later, the Academies issued a white paper on biotechnology, and two years after that produced a report on the field testing of genetically modified crops.

Given the tremendous growth in the number of transgenic products in today's marketplace, along with the public concern surrounding these products, the Academies launched a new study last year to review the current government system to regulate transgenic pest-protected plants and make suggestions for improvement, where improvements might be warranted.

This report is the result of that initiative. The project was funded entirely by the National Research Council and took 12 months to complete. Committee members were selected for their expertise in a number of areas including biology, agriculture, ecology, and the regulatory process.

Considering the current public debate over genetically modified organisms, we were not surprised that there would be intense scrutiny of our committee's composition. Clearly, this issue is a contentious one, with extraordinarily strong feelings on all sides. Nevertheless, our committee came together in an incredibly productive way, and worked very hard to examine the science and draw conclusions and recommendations based on science. The report represents a consensus among committee members who come from truly diverse perspectives and, as I've already mentioned, various areas of expertise. I am particularly proud to have been chair of this committee, and am equally, if not more, proud of the outcome: a strong framework for the future of regulation and research on pest-protected plants.

I must make a point of underscoring what we did and did not look at in our study. When I say transgenic pest-protected plants, I mean specifically plants whose genes have been modified through modern genetic engineering techniques, such as recombinant DNA technology, to express traits that make them resistant to certain pests and disease. These transgenic plants may include genes from distantly related species or even from different biological kingdoms. We did not look at plants genetically engineered for other purposes, such as to resist herbicides, or plants bred by other methods, although many of our findings apply to other categories of plants.

Because of public concerns about the safety of our food supply, we placed more emphasis on potential risks of transgenic pest-protected plants than on potential benefits. And we did not address the philosophical and social issues surrounding the use of engineering in agriculture, food labeling, or international trade.

Farmers have been using conventional breeding practices, such as hybridization, to develop crops with desirable traits for hundreds of years. Transgenic plants have only been grown commercially since 1995, although their use has increased dramatically since then. In 1999 alone, more than 70 million acres of transgenic crops were planted in the United States.

Given this striking increase in the number and types of transgenic plants on the market, our committee strongly believes that the federal agencies responsible for regulating them must take steps to better coordinate their work and to expand public access to the regulatory process. Public acceptance of these foods ultimately depends on the credibility of the testing and regulatory process, which must be as rigorous as possible and based on the soundest of science.

That said, I must also emphasize that the committee is not aware of any evidence suggesting that foods on the market today are unsafe to eat as a result of genetic modification.

Furthermore, we found no strict distinction between the health and environmental risks posed by plants modified through modern genetic engineering techniques and those modified by conventional breeding practices. In other words, the breeding process is not the issue; it is the product that should be the focal point of regulation and public scrutiny. That is, just because a plant is transgenic, doesn't make it dangerous.

This is why government regulation of these plants must continue to focus on their individual properties. To that end, our report makes several recommendations about research to improve what we know about these plants.

To date, only in very rare circumstances have pest-protected plants caused obvious health or environmental problems. For instance, although a human allergic reaction has never been documented for a commercially available transgenic plant, one such incident did occur at the research stage. In that study, people with a known allergy to Brazil nuts experienced a reaction when their skin was pricked with a solution from soybeans containing a Brazil nut gene. Thus, we believe high priority should be given to improving methods used to identify potential allergens, specifically focusing on new tests relevant to the human immune system, as well as to developing more reliable animal models.

We also believe that changes in physiology and biochemistry of pest-protected plants should be carefully monitored during development. And because the potential exists for transgenic plants to have increased levels of toxic compounds, the three agencies that regulate them should create a coordinated database to list information about natural plant compounds of dietary or toxicological concern. This would aid researchers who monitor concentrations of these compounds in such plants.

We also turned our attention to environmental concerns. We looked at the possibility that transgenic plants could inadvertently affect other organisms, for example, beneficial insects. As it turns out, both conventionally bred and transgenic pest-protected crops could impact these so-called non-target species, but the impact is likely to be smaller than that from chemical pesticides. In fact, when used in place of chemical pesticides, pest-protected crops could lead to greater biodiversity in some geographical areas. For that reason, we call for more research in this area.

You may recall the highly publicized report of monarch butterflies being affected by pollen from genetically engineered corn. This is a prime example of an issue that needs to be researched further, with rigorous field evaluations. In that particular paper, researchers reported that pollen from corn which had been genetically engineered to produce Bt toxins—a type of insecticide—slowed the growth of, and sometimes killed, the larvae of monarch caterpillars when enough pollen was placed on milkweed leaves fed to them in a laboratory. However, more recent studies suggest that pollen density in the field might be too low to pose a threat to the butterflies. Clearly, follow-up studies are needed in the field where pollen density might be lower and the toxin might be deactivated by environmental factors.

Concern also surrounds the possibility that genes for resisting pests might be passed from cultivated crops to their weedy relatives, potentially making weed problems worse. This could pose a high cost for farmers and threaten the ecosystem. We recommend further research to identify plants with weedy relatives, assess rates at which pest-resistance genes might spread, and develop techniques that would decrease this likelihood.

We also urge more targeted research to examine the potential for pests to evolve and develop a resistance to plants that have been genetically modified to kill them. Such resistance could result in a number of potential environmental and health consequences, including a return to the use of harmful chemical pesticides. We believe that strategies to manage the development of pest resistance should be encouraged for all types of a pesticide, be it in a spray form or produced by a plant.

At the core of these safety issues lies the federal system that regulates transgenic plants. Although the committee believes that generally the system is working well, we have identified need improvements. Our committee calls on the EPA, USDA, and FDA to improve the coordination of their regulation of these plants. This memorandum should identify regulatory issues under the jurisdiction of each agency as well as issues for which more than one agency has responsibility. It also should establish a process to ensure appropriate and timely exchange of information between agencies. For 14 years, the agencies have formulated policies for genetically modified foods under guidelines set forth in the 1986 Coordinated Framework for the Regulation of standards based on legal jurisdictions at the time. We believe that today the scope of each agency's oversight needs to be clarified, especially when a new product is to be reviewed by more than one agency.

Additionally, we believe the exemptions proposed in EPA's 1994 rule for regulating certain transgenic pest-protected plants need to be re-examined. EPA proposes to grant categorical exemptions for all plants that have been given a new gene from a sexually compatible plant, and for plants expressing proteins that are derived from a virus, known as viral-coat proteins. But in the first instance, we concluded that the transfer and manipulation of genes between sexually compatible plants could, in some cases, potentially increase human and environmental exposure to high levels of toxins. In the second instance, while plants with viral-coat proteins may be safe to eat, there are environmental issues to consider because of their potential to crossbreed with weedy relatives. We urge EPA to reconsider its plans to grant these categorical exemptions for transgenic plants.

Finally, we recommend that the agencies monitor ecological impacts of pest-protected crops on a long-term basis to detect any problems that may not have been predicted from tests conducted during the registration and approval process. And we call for a more open and accessible regulatory process to help the public understand the benefits and risks associated with transgenic pest-protected plants.

This concludes my opening statement. My colleagues and I will now take questions from credentialed reporters, alternating between questions in the room and those asked over the telephone. This news conference is being taped, so please step to a microphone to ask your question, and give your name and affiliation. Because this is a news conference, we ask that questions come from members of the news media only.

Thank you.

American Council on Science and Health

EDITORIAL

March 14, 2001

Biotech Detractors Distort Science to Support Their Views

by Ruth Kava

Opponents of foods altered by bioengineering (often misnamed genetically modified or GM foods) cite a number of concerns ranging from human safety to environmental degradation to support their disapproval of the technology. A new twist has surfaced recently: opponents have stated that the new golden rice, enhanced by the addition of genes that allow it to make beta-carotene, will not be effective at treating the vitamin A deficiency that is a major cause of childhood blindness in the developing world. They back up this claim by citing a number of supposed scientific "facts" about beta-carotene, and misinterpreting their meaning.

Good examples of such arguments are displayed in an article in the March 4 New York Times Sunday Magazine. The essay, by Michael Pollan, makes several faulty assertions about beta-carotene and golden rice. First, Mr. Pollan asserts that "an 11-year-old would have to eat 15 pounds of cooked golden rice a day to satisfy his minimum daily requirement of vitamin A." The fault here is that his so-called "minimum daily requirement" has no real meaning. This is a food labeling term that has not been used for many years.

Perhaps Mr. Pollan really meant the current Recommended Dietary Allowance (RDA) for vitamin A. But no knowledgeable nutritionist would assert that one has to consume the RDA of a vitamin daily to avert a deficiency disease. The RDAs (set by the Food and Nutrition Board of the National Academy of Sciences) are designed to incorporate safety factors to take into account individual variability in nutritional needs. They certainly do not constitute minimum requirements. For example, one could ward off scurvy (vitamin C deficiency) by consuming as little as 10 milligrams of the (missing line??) however, has recently been set at 90 milligrams per day—a substantial safety factor. These safety factors are included in all RDAs—the size will vary from one nutrient to another. Thus stating or implying that an intake of a vitamin at less than RDA levels, especially for one like vitamin A that is stored in the body, would have no beneficial effect on a deficiency condition is either a misunderstanding or misrepresentation of the facts.

A second assertion in Mr. Pollan's article is that malnourished children wouldn't be able to use the beta-carotene from golden rice because the body needs both fat and protein to change this nutrient to vitamin A. This is a tricky one, because to a limited extent, it is accurate. But what is omitted here is also important—that is, the body must also have fat and protein to absorb, store and use vitamin A. So if supplementation with beta-carotene would be ineffective because the children lacked fat and protein, it is likely that supplementation with vitamin A would be ineffective as well. Yet Mr. Pollan advocates "handing out vitamin-A supplements to children so severely malnourished their bodies can't metabolize beta-carotene."

In truth, a child who is deficient in vitamin A may well not be deficient in all nutrients, and vitamin A supplementation programs certainly can help. And so could additional intake of beta-carotene. Golden rice is not the only solution to the problem of vitamin A deficiency-blindness, but it certainly can be part of the solution.

There are many issues—scientific, economic, cultural, and political—that must be faced in order to deal with malnutrition of any type in the developing world. It is unlikely that any one approach will provide THE solution. We should not miss the chance to help fight one of the major causes of childhood blindness because of faulty "scientific" reasoning!

Ruth Kava, Ph.D., R.D., Nutrition Director, American Council on Science and Health

From Kava, R., "Biotech Detractors Distort Science to Support Their Views," American Council on Science and Health, March 14, 2001. With permission.

Killer Corn

Pesticide from engineered crop threatens butterflies —a consequence UCS foresaw

By Jane Rissler

My four-year-old friend Devin waved goodbye to his first monarch butterfly this spring. He and his preschool classmates had watched the monarch grow from a tiny caterpillar into a big, gorgeous orange-and-black-winged butterfly. Most American children, at least once during their school years, witness the majesty of the monarch.

No other insect is so precious to Americans—children *and* adults. It was not surprising then that a media firestorm erupted in May when Cornell University scientists reported that genetically engineered crops might endanger the butterfly. Americans took notice, as never before, of the risks of the new genetic technology that has moved quietly, with only meager government oversight, onto millions of acres.

Engineered Corn—A Threat to Monarchs?

In laboratory experiments, Cornell's John Losey found that pollen from Bt corn killed monarch butterfly caterpillars. The transgenic corn, engineered to kill pests like the European corn borer, produces an insect poison originally obtained from the soil bacteria *Bacillus thuringiensis* (Bt). Pollen from most Bt-corn varieties contains the Bt toxin.

In Losey's study, nearly half the caterpillars that ate milkweed leaves dusted with Bt-corn pollen died after four days, compared with no deaths among caterpillars eating leaves with normal corn pollen or no pollen at all. Bt-corn pollen also altered the eating behavior of the caterpillars that survived—they consumed far less and, as a result, grew much more slowly than those exposed to normal pollen.

We don't know yet whether monarch populations will be affected by the toxic pollen in their natural habitats—field studies will be required to determine if they are at risk. So far, however, results from a preliminary field trial at Iowa State University are not encouraging. Nearly a fifth of the monarch caterpillars placed on milkweed leaves taken from within and near Bt-corn fields died after only two days. By contrast, no caterpillars died on leaves with normal corn pollen.

Monarchs may be exposed to Bt-corn pollen during their annual migrations in the United States. About half of the eastern population of monarchs moves through the Corn Belt each spring and summer on their way from their winter habitat in Mexico to southern Canada—mating and reproducing as they go. Each generation produces caterpillars, which exclusively eat milkweed, a plant found along roadsides and in fields and pastures. To the extent that Bt-corn fields produce pollen at the same time that monarchs are in the area, caterpillars on milkweed in or near those fields may eat a deadly dose of Bt. Corn pollen can be carried hundreds of feet by the wind.

Other moth and butterfly caterpillars in the vicinity of Bt-corn fields may also be at risk, if they eat plants coated with toxic pollen. Scientists are just beginning studies to determine the extent of the threat to these and other insects. It's not yet known, for example, whether any of the 19 species of butterflies and moths listed as endangered or threatened under the Endangered Species Act are at risk from Bt-corn pollen.

Botched Risk Assessment

Why didn't the Environmental Protection Agency (EPA) and the US Department of Agriculture (USDA), both of which reviewed the environmental risks of Bt corn, uncover the threat to monarchs years ago—*before* they gave the go-ahead to sell Bt-corn seeds?

It should have been a simple matter to figure out that Bt-corn pollen is lethal to monarchs. Government scientists knew that the toxin was engineered into corn for the very reason that *it is specific to moths and butterflies*. The engineering was undertaken to give corn the ability to kill pest caterpillars, like the European corn borer and the corn ear worm.

In *The Ecological Risks of Engineered Crops* (1996), UCS predicted that pesticidal plants might harm insects other than the target pests and urged regulatory agencies to look at effects on relatives of the targets. In the case of the Bt corn, that would have meant testing the pollen on nonpest relatives of the European corn borer—butterflies and moths. However, according to government documents neither the EPA nor the USDA required companies registering the seeds to test either the toxin itself or the toxic pollen against any non-target butterflies or moths.

To make matters worse, the government apparently did not consider that butterflies and moths like the monarch, which do not eat corn, might nonetheless be exposed to the toxin in wind-blown pollen falling on plants that they eat.

The inability to recognize an obvious risk underscores what UCS and many other groups have been saying for years: US regulation of biotechnology is weak and not up to the task of protecting against its dangers. If regulatory agencies failed to see such a clear risk, how can they be trusted to figure out more complicated situations?

Questionable Benefits

Some level of risk with Bt corn might be acceptable if it offered substantial environmental benefits. A major selling point for Bt crops is their promise to reduce applications of chemical pesticides. So far, however, it appears that Bt corn may not fulfill that promise because farmers do not typically use pesticides to kill corn borers on field corn, to which the vast majority of corn acreage is planted in the United States. While Bt corn can increase yields, substituting the genetically engineered variety for regular corn may do little to reduce synthetic insecticide use.

What Next?

So far, the EPA—the agency with primary responsibility for regulating Bt corn—has had little to say about the monarch threat except that it is studying the matter. Fearing that the agency will not move aggressively to assure the safety of monarchs and other butterflies, UCS, along with other groups, is urging the EPA to upgrade its program for assessing the environmental risks of crops like Bt corn. Until that is done, we have asked that the EPA defer registration of new Bt-corn varieties or renewal of existing registrations. All current Bt-corn registrations expire by the spring of 2001.

Jane Rissler is a senior staff scientist in UCS's Agriculture and Biotechnology Program.

From Rissler, J., "Killer Corn: Pesticide from Engineered Crop Threatens Butterflies — a Consequence UCS Foresaw," *Nucleus*. With permission.

If You're Not Concerned About GM Foods You Will Be After You Read This

What the National Academy of Science Says About Genetically Engineered Crops

Peter Montague is the director of the Environmental Research Foundation and writes a weekly column known as Rachel's Environment & Health Weekly.

The genetically-engineered-food industry may be spiraling downward. Last July, U.S. Secretary of Agriculture Dan Glickman—a big supporter of genetically engineered foods (aka: genetically modified foods)—began comparing agricultural biotechnology to nuclear power, a severely-wounded industry. (Medical biotechnology is a different industry and a different story because it is intentionally contained whereas agricultural biotech products are intentionally released into the natural environment.)

In Europe, genetically engineered food has to be labeled and few are buying it. As the New York Times reported two months ago, "In Europe, the public sentiment against genetically engineered [GE] food reached a ground swell so great that the cultivation and sale of such food has all but stopped." The Japanese government also requires GE foods to be labeled. Americans in overwhelming numbers (80 percent to 90 percent or more) have indicated they want GE foods labeled but the GE firms consider a label tantamount to a skull and crossbones and the Clinton/Gore administration has sided with the biotech corporations against the people. To be fair, there are no indications that a Republican president would take a different approach. The biotech firms have invested heavily in U.S. elections and the resulting government represents their interests at home just as it does abroad. On this issue, to an astonishing degree, the biotech firms are the government.

Since the early 1980s, biotech corporations have been planting their own people inside government agencies, which then created a regulatory structure so lax and permissive that biotech firms have been able to introduce new genetically modified foods into the nation's grocery stores at will. Then these same "regulators" have left government and taken highly-paid jobs with the biotech firms. It represents an extreme case of the "revolving door" syndrome.

The U.S. regulatory system for GE foods, which was created in 1986, is voluntary. The U.S. Department of Agriculture regulates genetically engineered plants and the U.S. Food and Drug Administration (FDA) regulates foods made from those plants. If any of the plants are, themselves, pesticidal then U.S. Environmental Protection Agency gets involved. But in no case has any long-term safety testing been done. As the New York Times reported last July, "Mr. Glickman [U.S. Secretary of Agriculture] acknowledged that none of the agencies responsible for the safety of genetically modified foods—the Agriculture Department, the FDA, and the Environmental Protection Agency—had enough staff or resources to conduct such testing." At the time Mr. Glickman made his statement, 70 million acres in the U.S. had already been planted with genetically modified crops and two-thirds of the food in U.S. grocery stores contained genetically modified plant materials.

The importance of safety testing was emphasized by the National Academy of Sciences (NAS) in its latest (April 2000) report on biotech foods. The NAS said safety problems might include these:

- New allergens may be introduced into foods.
- New toxins may be introduced into foods. The NAS said "...there is reason to expect that organisms in US agroecosystems and humans could be exposed to new toxins when they associate with or eat these plants."
- Existing toxins in foods may reach new levels, or may be moved into edible portions of plants. ("Overall increases in the concentrations of secondary plant chemicals in the total plant might cause toxic chemicals that are normally present only in trace amounts in edible parts to be increased to the point where they pose a toxic hazard," NAS said.)
- New allergens may be introduced into pollen, then spread into the environment. [The NAS remains silent on the human-health implications of new allergens spread via pollen. If the biotech firms have their way, we will learn about this by trial and error. Unfortunately, trial and error has a serious drawback in this instance: once new genetic materials are released into the environment, they cannot be retrieved. Unlike chemical contamination, biotech contamination is irreversible.]
- Previously unknown protein combinations now being produced in plants might have unforeseen effects when new genes are introduced into the plants.
- Nutritional content of a plant may be diminished.

The mechanism for creating unexpected proteins or unexpected toxins or allergens would be pleiotropy, the NAS explained. Pleiotropy is the creation of multiple effects within an organism by adding a single new gene. In other words, putting a new gene into a tomato, intending to make the tomato more resistant to cold weather, might by chance, and quite unexpectedly, make some people allergic to the new tomato. "Such pleiotropic effects are sometimes difficult to predict," the NAS said. The NAS said that FDA, USDA and EPA all need to pay attention to such "unintended compositional changes" of genetically modified foods.

Unfortunately, as the NAS pointed out, current tests are not adequate for determining all the problems that might occur because of pleiotropic effects. For example if a new protein is created that has not previously been found in the food supply, then there is no reliable basis for predicting whether it may cause allergic reactions. Allergic reactions are not a trivial matter, the NAS pointed out: "...food allergy is relatively common and can have numerous clinical manifestations, some of which are serious and life-threatening."

New tests should be developed to test for allergenicity of genetically modified foods, the NAS said several times (the NAS called such new tests "highly desirable"). Specifically, the NAS recommended that tests be developed that actually measure reactions of the human immune system, which is the human system in which allergic reactions develop. The genetically modified foods on the market today have not undergone controlled experiments on real human immune systems. (Putting such foods into grocery stores is an uncontrolled experiment of sorts, but with no one collecting the data.)

In addition to human health problems, the NAS report discussed some of the agricultural and environmental problems that might occur from genetically modified (GM) plants:

- New chemicals in GM plants might kill predators and parasites of insect pests, thus leading to the loss of nature's own biological controls on certain pests.
- Plants themselves might become toxic to animals.
- Fallen leaves from GM plants might change the biological composition of the soil, leading to changes in nutrient uptake into plants or even toxicity to creatures living in the soil.
- Genes from genetically-engineered plants will escape and enter into wild species. This is called gene flow and the NAS says, "[T]otal containment of crop genes is not considered to be feasible when seeds are distributed and grown on a commercial scale." In other words, gene flow is going to occur. Wild plants are going to receive genes from genetically modified organisms. The biotech firms are re-engineering nature without understanding the means or the ends.
- When a plant is genetically engineered so that the plant itself becomes pesticidal (for example, Bt-containing corn, potatoes and other crops now planted on tens of millions of acres in the US), there may be effects on non-target organisms. In other words, pesticidal crops may affect creatures besides the specific pest they were intended to kill. The NAS says, "Non-target effects are often unknown or difficult to predict."

In sum, agricultural biotechnology has raced ahead at lightning speed (going from zero acres planted with GE crops in 1994 to 70 million acres planted in 1999) without any long-term testing, and with minimal understanding of the consequences. The NAS refers to these politely as "uncertainties" and it acknowledges that these uncertainties "often force agencies to base their decisions on minimal data sets."

So two-thirds of the food in US grocery stores contains plant materials that were genetically engineered. If they were subjected to government approval at all, it was on a strictly voluntary basis, and the government "often" approved new plants and new foods based on "minimal data sets," according to the National Academy of Sciences. Some of the most important aspects of these new foods had to be ignored because there is no way at present to test for them.

In sum, the biotech industry and its acolytes in government are flying blind and we are all unwitting passengers in their rickety plane. This is not a historical record that inspires confidence. No wonder the Clinton/Gore administration and the biotech corporations do not want anyone to know which foods have been genetically engineered. None of the biotech firms are even claiming that there are taste or nutritional benefits in the biotech foods being sold today, so, to put it bluntly, consumers would have to be out of their minds to eat this stuff or serve it to their children.

Given the serious problems that the NAS said may occur as thousands of new genetically modified foods are introduced into the US food supply without labels, naturally one wonders about liability insurance for the biotech industry. You will not find liability insurance discussed on the biotech industry's web site, www.whybiotech.com, so it is probably one of the industry's most serious problems.

Recently the Swiss company, Swiss Re, issued a report on GE foods. Swiss Re is a re-insurance company—it insures insurance companies against catastrophic loss. Swiss Re said genetic engineering "represents a particularly exposed long-term risk" and "genetic engineering losses are the kind which have not yet, or only rarely occurred and whose consequences are extremely difficult to predict."

Swiss Re then asked (and answered) the question, "…so how can genetic engineering risks be insured?" Here is Swiss Re's answer:

"It is currently not possible to give a direct answer to this question. A lot depends on whether consensus can be reached on the relevant loss scenarios in a dialogue involving the genetic engineering industry, society, and the insurance industry. This will make genetic engineering risks more calculable and more interesting to traditional insurance models. From the point of view of the insurance industry, we are at present a long way off. [Emphasis added.]

"Today we must assume that the one-sided acceptance of incalculable risks means that any participants in this insurance market run the risk not only of suffering heavy losses, but also of losing control over their exposure."

Without intending to do so, the Swiss Re report brings to mind an agenda for citizens who oppose the expansion of ag biotech:

On the principle that the polluter shall pay, biotech firms should be held strictly liable for any harms they may cause, not requiring proof of negligence;

Ag biotech corporations should not be allowed to self-insure; as we know from the asbestos industry, self-insurance can lead to bankruptcy and hundreds of thousands of legitimate claims never being paid;

Lawsuits should seek damages for gene flow, pollen drift, inadequate testing for allergenicity, crop failures, and so on. A series of lawsuits against private firms or government agencies would get the insurance industry's attention.

Stockholders in ag biotech firms should express concern (to the board of directors, and to the Securities and Exchange Commission) about the failure to disclose incalculable risks. Stockholders in insurance companies should express concern about the potential for "heavy losses" and "losing control over their exposure" if coverage is extended to ag biotech firms.

Montague vs. Stott

Peter Montague and Philip Stott.
Mr. Montague is the director of the Environmental Research Foundation.
Mr. Stott is a professor of biogeography at the University of London, UK.

Editor's Note: The following is a "TomPaine.com Face-off" between Professor Philip Stott and Peter Montague, on genetically modified foods. Professor Stott wrote to TomPaine.com in response to Mr. Montague's article, "If You're Not Concerned About GM Foods You Will Be After You Read This." Mr. Montague's piece originally appeared on Rachel's Environment and Health Weekly.

Professor Stott: Dear Tom Paine,

You are rightly admired for your 'common sense'; I hope you will not mind, therefore, if a fellow Brit pours scorn on the gross misuse of the National [Academy of] Sciences report [on GM Food] by Peter Montague in his travesty of an article on GM crops. The report went out of its way to say that the panel supported the development of these crops. How on earth can Mr. Montague write what he does?

Montague: So far as I know the National Academy of Sciences has never opposed any new technology. However, over the years, the Academy has expressed concerns about several new technologies, one of them being genetically engineered foods. In my article, I accurately reported the Academy's most recent expressions of concern about such foods.

Stott: The reality behind all the hysteria is that:-

Montague: As Professor Stott doubtless knows, hysteria has a specific meaning. I am not aware of any hysteria. Why does Professor Stott fail to offer evidence for his assertions?

Stott: (a) biotech crops are already the most tested ever (nine years to commercialization) — cf. this with all the totally untested so-called 'health foods' littering the shelves of pharmacies and chemists;

Montague: Now Professor Stott offers us a red herring. The National Academy didn't ask, "Have genetically engineered foods been tested more or less than some other foods?" The Academy asked, "Have genetically engineered foods been tested sufficiently to protect public health and the environment?" In answering that question the Academy criticized the safety protocols that have been used thus far, and the Academy said it is "highly desirable" that better safety tests be developed and used. Evidently Professor Stott wants to divert our attention away from the Academy's important conclusion that genetically engineered foods have been inadequately tested for safety.

Stott: (b) During the whole of their now long development and use, there hasn't been a single case of a problem with nutrition (and don't bring up the L-tryptophan business—that was discredited ages ago) — cf. this with E-coli and organic production;

Montague: Again, Professor Stott is stating conclusions not based on any facts. How can Professor Stott know there have been no problems? The National Academy said repeatedly that genetically modified foods might cause allergic reactions in some people and that allergic reactions can be serious, even life-threatening. To know whether allergic reactions were occurring, one would need to maintain a registry of allergic reactions and then examine individual cases to pin down the cause. No such registry exists. Physicians in this country are not required to report allergic reactions that they observe in their patients. Therefore, relevant data are not being collected, much less analyzed. Refusing to look for a problem is not the same thing as having evidence that a problem does not exist. The truth is, no one knows the consequences of exposing hundreds of millions of people to genetically engineered foods, and, furthermore, no one it making any effort to find out. Thus Professor Stott's conclusion is not based on data. To put it bluntly, he made it up.

Stott: (c) There is growing evidence that the consequent reduction in chemical spraying is already increasing, for example, bird biodiversity—and just look, by contrast, at the total misuse of the Monarch butterfly claims, which were shown to be ecologically meaningless. In the USA alone, there have been over 5,000 trials and 24,000 field trials!

Montague: On two counts, Professor Stott appears to be badly misinformed. First, there is solid evidence from farms in this country that genetically modified crops often require more pesticides than traditional crops, not less. The corporations that sell genetically modified seeds initially claimed that such crops would reduce pesticide use, but actual experience on American farms has proven these claims false. Why is Professor Stott repeating these false claims? (To get the facts, read http://www.pmac.net/IWFS.pdf or listen to http://www.aces.uiuc.edu/worldfood/1999/broadcase/schedule.html).

Secondly, the National Academy of Sciences says that damage to non-target organisms (such as Monarch butterflies) is poorly understood and that "further field-based research is needed" before conclusions can be drawn. (See page 38 of the Academy's report, available at http://www.nap.edu/html/gmpp/ Professor Stott has stated a conclusion that the Academy said cannot be supported by available facts.

Stott: Frankly, the last thing you want is a cross-fertilization of America by the current hysteria from Europe. The reality is that these are the crops of the future and it is vital that the USA leads the way. Your readers might like to visit the 'ProBiotech Web Site' (http://www.probiotech.fsnet.co.uk) for a full refutation of Mr. Montague's gross distortions.

Montague: If these are the foods of the future, as Professor Stott claims they are, why not label them and let people decide for themselves what's best? If these products are so beneficial and safe, why are the biotech firms and the government afraid to label them, afraid to allow people to make informed choices in the grocery store?

In sum: It is evident that Professor Stott has abandoned his role as a serious scholar and has become a cheerleader for the biotech industry. Frankly, his willingness to disregard the available facts makes me wonder whether the good professor has been taking research money from the biotech corporations.

Professor Stott Fights Back:

Thanks for asking me to reply to Peter Montague's comments on my brief critique of his original article on biotech crops.

You always know that an argument is 'weak' when the protagonist has to resort to ad personam abuse. Peter writes: "It is evident that Professor Stott has abandoned his role as a serious scholar and has become a cheerleader for the biotech industry. Frankly, his willingness to disregard the available facts makes me wonder whether the good professor has been taking research money from the biotech corporations."

This could not be farther from the truth. I have no financial links whatsoever with any biotech company; I do not even have shares in any such company—and quite deliberately so, precisely because I wish to maintain my academic independence at all times. My 'ProBiotech' Web Site is run completely from my own pocker (at, I may add, c. 300 dollars per month and to my wife's chagrin) and I write all my own HTML and JavaScript. I find it very depressing that Peter cannot accept that the 'good professor', as he terms me, is not allowed to reach an honest conclusion that just happens to differ markedly from his own.

Peter then wishes to be pedantic about my use of my word 'hysteria'; I am happy to replace this with 'ecochondria' or 'ecohype' if he would prefer?

And hype it all is. The fact is that biotech plants have been around since 1983, the first commercial product since the late 1980s, and the first full product, the famous FLAVR SAVR **TM** tomato, since 1994. Moreover, between 1996 and 1999, the world commercial acreage of biotechnology crops rose from 4.3 to over 100 million acres, with 81 million acres in Canada and the US, 16 million acres in Argentina, and 1 million acres in both China and Australia. The 1999 Canada/US figure included 28.3 million acres of corn, 35 million acres of soybeans, and 5.3 million acres of oilseed rape (canola). In 2000, 52% of the US soybean crop is predicted to be genetically modified.

And throughout all this long history there has not been one single substantiated problem with either human nutrition or the environment. As the National Academy of Sciences says: "The committee is not aware of any evidence that foods on the market are unsafe to eat as a result of genetic modicitaion." That is all anyone can ever say. 100% certainty is never an option in anything. And, with regard to the environment, in the US alone there have been over 5,000 full trials and over 24,000 field trials.

Now, of course, more research, more trials, more care is always needed. But I can publically state that I, for one, would be much happier eating a biotech product than either a conventional product or, for that matter, a so-called 'organic' product. Moreover I am all for labelling, but **only** if that label is truly informative, without bias, and legal. Coming up with labels that are truly informative and unbiased is far from easy. Do you reflect process, protein product, both, or what? No label must indicate either advertently or inadvertently that something is 'better' or 'worse' without clear evidence to that effect. Already US attempts to define 'organic' are running into precisely this problem.

To sum up therefore: biotech crops are the most tested to date; we must, of course, always test more, and we can label if it is truly informative, unbiased and legal. I am not sure, however, that this is possible in the climate currently being engendered by Peter and his extreme environmentalist friends.

Thanks again for letting me come back on this.

Philip Stott

Peter Montague's Very Last Jab:

Professor Stott writes, "But I can publically state that I, for one, would be much happier eating a biotech product than either a conventional product or, for that matter, a so-called 'organic' product."

This reminds me of the advocates for nuclear power who used to offer to eat plutonium to show how safe it really is. I always thought it was an excellent idea for nuclear power advocates to eat plutonium, the more the better.

And for the same reason I think it is an excellent idea for Professor Stott to eat as much genetically modified food as he can get his hands on. (I would, however, prefer that this food be grown in a laboratory so that Professor Stott doesn't irretrievably contaminate the nature environment with rogue gene products while he fulfills his kinky fantasy.)

Then we all relax and let Darwinian evolution take its course.

Risks and Precaution in Agricultural Biotechnology

A Role for Science and Scientists

Katherine Barrett, Ph.D.[1]

The precautionary principle has existed as a formal decision making process for over 30 years, and has figured largely in recent debates on genetically engineered (GE) organisms. For example, the European Union supported a precautionary approach in 1999 by instituting a de facto moratorium on the commercialisation of GE organisms. In January 2000, precaution was incorporated into the international Cartagena Biosafety Protocol on the transfer of "living modified organisms." These moves have not only raised the prominence of the precautionary principle, but have also highlighted an urgent need to examine relationships among risk assessment, precaution, and current regulations on plant biotechnology safety and trade.

This paper looks at interpretations of the precautionary principle, how this principle might be applied to the potential hazards of plant biotechnology, and what a precautionary approach would mean for scientific research and the scientific community. I will base my discussion largely on current policies and regulations in Canada although many of my points also apply more widely.

What is the Precautionary Principle?

In general, the precautionary principle is a process for decision-making under conditions of uncertainty. The principle states that when there is reason to believe that our actions will result in significant harm, we should take active measures to prevent such harm, even if cause-and-effect relationships have not been proven conclusively. Precaution first arose as a formal principle for environmental policy in Germany in the 1970s. It has since been invoked in many international laws, treaties and declarations on a range of environmental issues including climate change, marine dumping of pollutants, and general efforts towards sustainability. With its incorporation into the Rio Declaration of 1992 and the Biosafety Protocol in 2000, there is growing consensus that the precautionary principle has reached the status of international customary law.

To date, formulations of the precautionary principle have ranged from strong to quite week. Despite this variability, there are several consistent core elements of precaution:

---

[1] Currently project director with the Science and Environmental Health Network, and Research Associate with the Eco-Research Chair of Environmental Law and Policy, University of Victoria, Canada. This paper was originally presented at the International Botanical Congress, in St. Louis MO, August 1999.

1. *Protection of the environment*
   The overall aim of the principle is to protect the environment from negative effects of human activities. Many interpretations further specify that we should avoid harms that are irreversible, persistent, bioaccumulative or otherwise 'serious'.
2. *Proactive, anticipatory action*
   Most versions of the precautionary principle stress the need for proactive measures to anticipate and prevent potential harms at the source, rather than reactive measures which aim to mitigate, eliminate or compensate for harm once it has occurred. The precautionary principle is therefore fundamentally a research, knowledge and action oriented principle.
3. *Recognition of uncertainty*
   However, the precautionary principle also requires recognition of uncertainty and an appreciation of the limits of scientific knowledge. That is, we cannot always expect a full and conclusive understanding of potential environmental effects, especially when such effects are long-term, unconfined and broad-scale. More importantly, in many cases we simply cannot afford to wait for conclusive proof of cause-and-effect relationships before taking action to prevent harm.
4. *Shifting the burden of proof*
   The precautionary principle advocates shifting the burden of proof to the developers of potentially hazardous technologies. This is a particularly difficult and contentious element of the principle and has been interpreted in several ways. Basically, it requires proponents to demonstrate that the technology in question does not pose unreasonable harm and that there are no less-damaging alternatives. The rationale is that those who stand to gain from an activity ought to assume the responsibility and costs of demonstrating its safety.
5. *Cost effectiveness*
   Finally, some versions of the precautionary principle include analysis of the costs and benefits of precautionary action. For example, the Rio Declarations states: "Where there are threats of serious or irreversible damage, lack of full scientific certainty shall not be used as a reason for postponing *cost-effective* measures to prevent environmental degradation" (emphasis added). Stronger interpretations of precaution question whether trade-offs between economic benefits and environmental protection are necessary, and place greater emphasis on developing safer alternatives to potentially hazardous activities. While in practice (and in politics), it is often difficult to completely ignore measures of costs and benefits, it is generally accepted that if cost-benefit analysis is made consistent with a precautionary approach, it must be weighted to include non-monetary elements such as societal values, dangers imposed on future generations, harm to non-human beings, and recognition of uncertainty and ignorance. In this sense, cost-benefit analysis is *one tool* under a precautionary umbrella, but is not a sufficiently robust decision-making framework to stand alone.

As this brief description suggests, the precaution is an evolving principle of environmental law, policy and ethics. The current flexibility in interpretation and implementation is not unusual: customary law, as well as current 'case-by-case' regulations for plant biotechnology, are refined over time through application. Some commentators on the precautionary principle hold that it is necessarily a general principle and that specific applications will require secondary legislation appropriate to the case.

Consistent interpretation is only of several hurdles to implementing the precautionary principle. More difficult will be clarifying the relationship between precaution and scientific research. Some critics have claimed that the precautionary principle is inherently non-scientific or blatantly anti-scientific because it advises taking preventative action prior to conclusive scientific proof of harm. In contrast, I would like to emphasize that the precautionary principle does *not* aim to stifle discovery or shut down research. Just the opposite. The precautionary principle requires *more* research in order to understand and forestall the possible adverse effects of technologies. Perhaps the difference in these interpretations lies in the timing: a precautionary approach entails in-depth research into a broad range of potential impacts, and development of effective safeguard *prior to*—or at the very least simultaneously with—commitment to new technologies. The principle therefore points to a vital and active role for the scientific community: a broader research agenda and more explicit social responsibilities. How might this interpretation of the precautionary principle be applied to plant biotechnology?

Impacts of Plant Biotechnology

The rate of adoption of GE crops has been staggering. The first small-scale field trials were conducted in the mid 1980s. In 2000, almost 45 million hectares of GE crops were planted world-wide, with the U.S., Argentina and Canada accounting for 99% of the total acreage. Supported by over twenty years of research and development and firm commitment by industry and government, GE food is now firmly integrated into the North American agricultural system.

What hazards are associated with the crops? Several recently published experiments suggest that GE crops may have adverse, unexpected effects on the environment and human health. While controversy and conflicting interpretations surrounds some of these studies, I suggest that such experiments provide an excellent indication of the current state of knowledge regarding the risks of plant biotechnology: yes, there is *evidence* of potentially serious hazards; but no, we do not have absolute proof of risk *or* safety.

What is clear, however, is that these issues are confounded not only by scientific uncertainty but also by long-term political commitments, huge economic stakes, and profoundly conflicting ethical principles. Such social dimensions of risk may actually contribute to the overall uncertainty surrounding plant biotechnology and to the complexity of decision-making. We must ask: What questions have not yet been explored—and why? What kind of research *is* being conducted—and by whom? How are the potential hazards of plant biotechnology being assessed and regulated?

If critics of the precautionary principle are concerned about lack of scientific rigour, it seems fair to ask whether current regulatory systems are in fact based on rigorous, 'sound' science. I would like to suggest several ways in which existing regulations may be compromised in this respect.[2]

- First, there is a prior presumption in current regulations that GE crops are not inherently different from traditionally bred crops. This basic premise of product-based regulation tends to focus assessments at the phenotypic level, and it effectively places the burden of proof on those who claim that GE crops *do* in fact pose unique concerns that merit greater regulatory consideration.
- Second, the data used in risk assessments of GE crops are supplied almost entirely by the proponent of those crops. In Canada, much of this data is protected as confidential business information, and all of it must be obtained through our Access to Information and Privacy Act. There is no independent peer review or testing of the data. The final decision is made by the federal government who has been an advocate of biotechnology since the early 1980s. In this system, we might say the "*benefit* of proof" already lies with the proponents.
- A third related point is that to date, risk assessments for GE crops have been restricted to very narrow questions, limited methods for addressing those questions, and short time frames in which to observe effects. For example, testing usually takes place over the course of 1 or 2 field seasons, and in confined field trials or during pre-commercial variety trials. Consequently little to no emphasis is placed on long-term, indirect, or *unexpected* effects and very broad conclusions are drawn from a limited data set. As several Canadian scientists point out in a recent editorial on this issue: "If you don't look for something, you are rather unlikely to find it."
- Finally, while the benefits of plant biotechnology have long been claimed, they have only recently been independently investigated—and inconclusively so. Whether currently marketed GE crops will increase yields, reduce chemical input, and feed needy populations remain very much open questions, and will likely depend on local ecological, agronomic and political systems. Assessment of these claims is made more difficult by the relative lack of funding for alternatives such as low input and organic agriculture.

I suggest that we are a long way from implementing the precautionary principle or employing 'sound science' to assess the hazards of plant biotechnology. How can we move toward these goals?

Toward a Precautionary Approach to Plant Biotechnology

I will conclude by sketching a type of science which aims to confront the challenges of environmental protection and technological development "head on". A

[2] These comments are based on analysis of the Canadian regulatory system conducted by K. Barrett as part of her doctoral dissertation: Canadian Agricultural Biotechnology: Risk Assessment and the Precautionary Principle. 1999. Department of Botany. University of British Columbia. Vancouver.

"precautionary science" provides a rigorous foundation for assessing the potential hazards of plant biotechnology *and* a research program that is fully consistent with the precautionary principle.

- First, a precautionary science broadens the scope and scale of inquiry. What types of hazards are we looking for? How can we actively investigate the long-term, indirect, cumulative and synergistic effects of plant biotechnology? How can we include social and ethical aspects in our investigation? Broadening the scope even further, we might step back and ask, what is the central problem? Is, in fact, precise measurement of the risks of plant biotechnology our first priority? Or are there more *direct* ways of tackling problems such as chemical-based agriculture and the world's food needs—issues which plant claims to address? What alternatives do we have or should we pursue?
- Second, precautionary science recognizes that uncertainty extends beyond technical hurdles or a temporary lack of data. A more profound type of uncertainty arises from the discrepancy between the necessary limits of experimental research and the indeterminate world in which results are interpreted and applied. This kind of uncertainty is largely irreducible—it is an inherent aspect of conducting science within complex ecological, social and political circumstances, and point to the limits of scientific knowledge and authority. Recognition of such contingencies need not stifle research. On the contrary it should *stimulate* creative and collaborative ways to address urgent and complex problems.
- Third, recognising that scientific research is not conducted in a social vacuum, and that complex environmental health issues cannot be solved in isolation, a precautionary approach to science is necessarily more inclusive and cooperative. This means greater collaboration among disciplines (extending beyond our natural sciences to include social sciences and humanities). It also means extending our community of "peers" to include a diversity of knowledge and experience and to better represent those people who may be adversely affected by technologies. Equally important, peer review should be extended to research conducted by the private sector. All of this points to a more engaged scientific community—one which is responsive to, and responsible for, the *public* interest. Interaction with "the public" must therefore extend beyond an educate-and-inform approach to more of a partnership which entails collaborative learning and problem-solving.
- Finally, I suggest that the scientific community must accept their share of the burden of proof. Definitions of adequate proof depend largely on how, and by whom, the terms of inquiry are framed. This burden cannot rest on the shoulders of one interested party. All parties are interested, and all have responsibilities. I invite the scientific community to accept an active, engaged and precautionary role in addressing issues of risk and ethics in plant biotechnology.

SUMMARY OF FRAMEWORK FOR APPLYING THE PRECAUTIONARY PRINCIPLES TO GE CROPS

Core Element of the Precautionary Principle	Step in Precautionary Framework	Specific Considerations
Protection of Public Health and the Environment	Setting Goals	• Articulate broad long-term goals for agriculture • Determine how agricultural biotechnology may further or hinder these goals • Articulate the problems that GE crops aim to solve • Identify the sources of these problems • Assess claims that GE technology is necessary to achieve goals and address problems
	Assessing Alternatives	• Research and assess a range of alternatives to genetic engineering • Include people with a diversity of interests and values in the assessment; ensure assessments open to public scrutiny and revision • Evaluate social and political factors that encourage or discourage alternatives
	Research to Support Long-Term Goals	• Ensure adequate funding and support for a broad range of agricultural technologies and practices • Ensure long-term, multi-disciplinary, and participatory studies on potential effects are well supported, funded and distributed
	Open Decision-making	• Establish open decision-making processes • Ensure these processes are timely, effective and binding • Ensure decisions are reversible and processes iterative • Ensure information is made freely available (i.e. not proprietary or confidential business information)
Identifying Potential Harm	Defining Harm	Consider: • Biological, ecological, social, economic and political effects • Direct, indirect, cumulative and synergistic effects • Immediate, delayed, persistent and long-term effects • Effects of cultural or geographic context • Whether commitment to agricultural biotechnology forecloses other options
	Defining Standards	• Identify standards against which agricultural biotechnology is assessed (e.g. industrial-scale, chemical-based agriculture) • Consider potential effects of these standards • Assess whether GE crops perpetuate potentially harmful trends in agriculture
	Distribution of Harm	• Evaluate distribution of hazards and benefits • Determine if agricultural biotechnology tends to concentrate or distribute authority (empowers or disempowers those who may be affected)

SUMMARY OF FRAMEWORK FOR APPLYING THE PRECAUTIONARY PRINCIPLES TO GE CROPS (continued)

Core Element of the Precautionary Principle	Step in Precautionary Framework	Specific Considerations
	Degree of Harm	• Evaluate evidence that harm is potentially irreversible, widespread, long-term, accumulative, persistent or toxic • Consider whether potential harms been accepted voluntarily or imposed • Evaluate links among physical, political, social, and psychological dimensions of harm
Recognition of Scientific Uncertainty	Types of Uncertainty	• Distinguish among situations of certainty, risk, uncertainty, ignorance and indeterminacy • Identify technical and methodological uncertainties • Articulate assumptions and limitations that contribute to "great uncertainty" • Evaluate ways in which these uncertainties bear on (qualify) research results; ensure this influence acknowledged • Assess means of reducing uncertainties • Consider who may benefit and who may be harmed from these uncertainties (e.g., if technology proceeds or is halted due to uncertainty)
	Evidence of Harm	Consider: • Accuracy. Experiments can be repeated with similar results under similar conditions • Validity. Evidence is relevant to 'real world' conditions (i.e., have uncertainties and local contingencies been acknowledged) • Sources. Identify sources of evidence (e.g., Have a number of people advanced or supported this claim; Are opinions divided along professional, disciplinary or political lines?) • Plausibility. Evidence seems reasonable or believable • Coherence. Relation of evidence to existing theories, data or case studies. Can the evidence be supported through diverse means, e.g. – quantitative and qualitative data – correlation, pattern, association – experimental data – experiential information (e.g. What can be learned from past experience with similar technologies? What are the experiences of those who use the technology?) – local, context-specific information (e.g. case studies) – general principles

SUMMARY OF FRAMEWORK FOR APPLYING THE PRECAUTIONARY PRINCIPLES TO GE CROPS (continued)

Core Element of the Precautionary Principle	Step in Precautionary Framework	Specific Considerations
	Research on Potential Hazards and Benefits	• Consider broad definitions of harm and temporal/spatial scales (see "Identifying Potential Harm" above) • Ensure conclusions are sensitive to false negatives as well as false positives, and error biases are made explicit • Calculate statistical power of the data (if conclusions are based on statistical analyses) • Determine if experiments and conclusions are context specific or conditional (e.g. Are results extrapolated to more general circumstances? How are effects of conditionality assessed and acknowledged?) • Assess potential benefits through similarly comprehensive and rigorous methods
	Conditional Approvals & Monitoring	• Establish appropriate and effective responsibility for monitoring • Specify which parameters will be monitored, and through what processes • Specify measures to be taken if adverse effects are detected • Specify measures to be taken if irreversible effects are detected
	Moratoria	• Continue appropriate and adequate research on the hazards of GE crops and alternative technologies
	Phase-Outs & Bans	• Ensure alternatives have been adequately investigated
Shifting of the Burden of Proof	Testing	• Ensure proponents assume due responsibility for testing environmental and health effects of GE crops • Ensure appropriate research methods (outlined above) been adopted
	Consultation & Review	• Establish independent review process for data, methods and conclusions of tests conducted by proponents • Ensure public disclosure of data prior to decision-making
	Notification & Informed Consent	• Establish processes for appropriate labelling, tracing, and segregation of GE • Ensure fair advertising practices • Establish processes for advanced informed agreement among exporters and importers
	Protective Measures	• Ensure proponents adopt all known protective measures
	Liability & Financial Responsibility	• Establish procedures for environmental bonds • Ensure proponents are held liable for adverse effects

Section V

Environmental Advocacy Documents

Primary sources produced and distributed by environmental advocacy organizations are becoming more rare, probably due to the increasing costs of doing so, but, perhaps as well because there is such a large pool of information and position papers already in circulation, especially on the Web, and there is less and less reason to generate new documents. New hard copy advocacy papers are confined principally to newsletters, newspaper ads, and extended press releases.

The first documents included here are from *Rachel's Environment and Health Weekly,* one of the most respected and quoted publications in the field. It is researched and written by Dr. Peter Montague, whose work is frequently circulated by TomPaine.com, an environmentally friendly website. The other hard copy reprinted here is a piece from *News on Earth*, published by The Public Concern Foundation.

TomPaine.com also publishes op-eds in the *New York Times,* one of which I have collected here. The Turning Point Project, "turnpoint.com," which has made the threats posed by biotechnology one of its principal targets, also publishes its work in an occasional series of *New York Times* op-ed pieces. The Campaign to Label Genetically Engineered Foods is, as its name suggests, a non-profit political advocacy organization dedicated exclusively to creating a national grassroots consumer effort to promote the enactment of a GM food labeling law. It communicates almost exclusively through its Web site, "thecampaign.com," but it distributes "Action Packets" in public places, most prominently with the Whole Foods chain, where it can expect a receptive audience. Reprinted here is the substantive portion of one of these packets.

The mainstream organizations, as suggested earlier, work principally through the Internet and their own newsletters. The Sierra Club convened a task force on genetic engineering, and has published its report on the Internet. It is included here, as is a piece from *Greenpeace Magazine*, (Winter, 1996). Reprinted here is a news brief from Worldwatch (February 17, 2000) and a "fact sheet" from Environmental Defense.

Finally, I must mention the work of Friends of the Earth, a major environmental advocacy group. While there are few if any brief documents available from it (it does publish full length books), it has constituted itself as a kind of clearinghouse. Its Web site, "foe.org," will lead one to worldwide press releases, lists of U.S. and foreign organizations opposed to genetically modified foods, lists of food manufacturers that routinely incorporate GMOs into their products and lists of foods with genetically processed ingredients, and notices of relevant public meetings. It calls this effort the "Safer Food, Safer Farms Campaign."

Discussion

The big question regarding the advocacy of environmental organizations, so much of which is conducted on the Internet, is, of course, its effectiveness. Working through Web sites limits their audiences to those who specifically call up their homepages, or (hopefully in their mind) journalists. But they can communicate at a fraction of the cost that would be involved in generating and circulating hard copy. How do the plusses and minuses of this transition play out? How often have any of us logged on to one of these sites, absent a specific reason to do so?

On the other hand, is it safe to say that it is much more likely that a *New York Times* reader will see a TomPaine or Turning Point Project op-ed piece, or pick up a "thecampaign" leaflet in the grocery store? Do Web sites commit the organizations to "preaching to the choir?"

Format aside, is there a detectable theme to the issues presented by the organizations in whatever form they take? Do they trade in particular kinds of concerns? Again, comparison with those in the next section on business and industry will prove interesting.

Environmental Research Foundation Home

Rachel's Environment & Health News

#695 - Biotech In Trouble--Part 1, May 04, 2000

Biotech In Trouble--Part 1

The agricultural biotechnology industry's situation is desperate and deteriorating. To be sure, genetically engineered (GE) food is still selling briskly on grocery shelves in the U.S. but probably only because GE products are not labeled, so consumers have no idea what they're buying.

At present, an estimated 2/3rds of all products for sale in U.S. grocery stores contain genetically engineered (GE) crops, none of which are labeled as such.[1] However, polls show that U.S. consumers overwhelmingly want GE foods labeled. In a TIME magazine poll in January, 1999, 81 percent of respondents said genetically engineered foods should be labeled.[2] A month earlier, a poll of U.S. consumers by the Swiss drug firm Novartis had found that more than 90% of the public wants labeling.[3] The NEW YORK TIMES reported late last year that a "biotech industry poll" showed that 93% of Americans want genetically engineered foods labeled.[4] Legislation requiring labels on GE foods was introduced into Congress last November by a bi-partisan group of 20 legislators.[5]

For five years the GE food industry has been saying GE foods couldn't be labeled because it would require segregating GE from non-GE crops -- a practical impossibility, they said. However, in December, 1999, Monsanto announced that it had developed a new strain of rapeseed (a crop used to make canola cooking oil) that might raise the levels of vitamin A in humans.[6] How could consumers identify (and pay a premium price for) such a product if it weren't labeled? Obviously labeling will become possible -- indeed, essential -- when it serves the interests of the biotech corporations.

Many food suppliers seem to have figured out for themselves how to segregate GE crops from non-GE. According to the NEW YORK TIMES, Kellogg's, Kraft Foods, McDonald's, Nestle USA, and Quaker Oats all sell gene-altered foods in the U.S. but not overseas.[7] Gerber and H.J. Heinz announced some time ago that they have managed to exclude genetically modified crops from their baby foods.

For its part, the U.S. government has steadfastly maintained that labeling of GE foods is not necessary -- and might even be misleading -- because traditional crops and GE crops are "substantially equivalent." For example, the government has maintained that Monsanto's "New Leaf" potato -- which has been genetically engineered to incorporate a pesticide into every cell in the potato, to kill potato beetles -- is substantially equivalent to normal potatoes, even though the New Leaf potato is, itself, required to be registered as a pesticide with U.S. Environmental Protection Agency (EPA). (See REHW #622.)

Now the government's position has become untenable. In February of this year, the government signed the international BioSafety Protocol, a treaty with 130 other nations, in which all signatories agree that genetically modified crops are significantly different from traditional crops. Thus with the swipe of a pen, the U.S. government has now formally acknowledged that GE crops are not "substantially equivalent" to traditional crops.

Meanwhile, a groundswell of consumer protest reached a crescendo last year in England and Europe, then spread to Japan and the U.S. where it has severely eroded investor confidence in the industry. Major U.S. firms that had invested heavily in the technology are now being forced to pull back. As we reported earlier (REHW #685), Monsanto, Novartis, and AstraZeneca all announced in early January that they are turning away from -- or abandoning entirely -- the concept of "life sciences" -- a business model that combines pharmaceuticals and agricultural products. The NEW YORK TIMES reported in January that American Home Products -- a pharmaceutical giant -- "has been looking for a way to unload its agricultural operations." At that time the TIMES also said, "Analysts have speculated that Monsanto will eventually shed its entire agricultural operation."[8] In late February, DuPont announced that it was returning to its traditional industrial chemical business to generate profits. The WALL STREET JOURNAL said February 23, "But the big plans DuPont announced for its pharmaceuticals and biotech divisions fizzled as consolidation changed the landscape, and investor enthusiasm cooled in the face of controversy over genetically engineered crops."[9]

Investors are not the only ones turning away from genetically engineered foods. The WALL STREET JOURNAL announced in late April that "fast-food chains such as McDonald's Corp. are quietly telling their french-fry suppliers to stop using" Monsanto's pesticidal New Leaf potato. "Virtually all the [fast food] chains have told us they prefer to take nongenetically modified potatoes," said a spokesperson for the J.M. Simplot Company of Boise, Idaho, a major potato supplier.[10] The JOURNAL also reported that Procter and Gamble, maker of Pringles potato chips, is phasing out Monsanto's pesticidal potato. And Frito-Lay --which markets Lay's and Ruffles brands of potato chips -- has reportedly asked its farmers not to plant Monsanto's GE potatoes. A spokesperson for Burger King told the WALL STREET JOURNAL that it is already using only traditional potato varieties. A spokesperson for Hardees, the restaurant chain, told the WALL STREET JOURNAL that Hardees is presently using Monsanto's pesticidal potato but is considering whether to abandon it.

Earlier this year, Frito Lay also told its corn farmers to abandon genetically-modified varieties of corn for use in Doritos, Tostitos, and Fritos.[7]

According to the NEW YORK TIMES, U.S. farmers have sustained a serious financial blow because they adopted genetically engineered crops so rapidly. In 1996, the U.S. sold $3 billion worth of corn and soybeans to Europe. Last year, those exports had shrunk to $1 billion -- a $2 billion loss. The seed sellers like Monsanto and DuPont got their money from the farmers, so it is the farmers who have taken the hit, not the ag biotech firms. [11]

The WALL STREET JOURNAL reported April 28 that, "American farmers, worried by the controversy, are retreating from the genetically modified seed they raced to embrace in the 1990s... government and industry surveys show that U.S. farmers plan to grow millions fewer acres of genetically modified corn, soybeans and cotton than they did last year."[10]

The ag biotech firms dispute this assessment. They say demand for genetically modified crops has never been better. Less than a year ago Robert Shapiro, the chief executive officer of Monsanto, said bravely, "This is the single most successful introduction of technology in the history of agriculture, including the plow."[12] This year a spokesperson for Monsanto says, "We're seeing a very stable market. There's no major step backward; it's now a matter of how much we'll grow." [11] But Gary Goldberg, president of the American Corn Growers Association, told the NEW YORK TIMES recently that he believes that genetically modified (GM) corn plantings will be down about 16% this year, compared to last. He indicated that the ag biotech firms are resorting to deception to maintain sales: "The [ag biotech] companies are deceiving farmers into thinking their neighbors are planting G.M.," he told the NEW YORK TIMES.[11]

In coming days, genetically engineered (GE) food is likely to get more attention from the public. Last month the National Academy of Sciences issued a report confirming what critics have been saying about GE crops: they have the potential to produce unexpected allergens and toxicants in food, and the potential to create far-reaching environmental effects, including harm to beneficial insects, the creation of super-weeds, and possibly adverse effects on soil organisms. The Academy said there was no firm evidence that GE foods on the market now have harmful effects on humans or the environment, but the Academy also indicated that testing procedures to date have been woefully deficient.[13] Indeed, the present regulatory system is voluntary, not mandatory, so it is possible that the government may not even know about all of the genetically engineered foods being sold in the U.S. today.

The Academy pointed out that roughly 40 GE food products have, so far, been approved for sale in the U.S. but approvals have also been given for an additional 6,700 field trials of genetically modified plants.[13,pg.35] And a NEW YORK TIMES story May 3 about super-fast-growing GE salmon noted that "a menagerie of other genetically modified animals is in the works.... Borrowing genes from various creatures and implanting them in others, scientists are creating fast-growing trout and catfish, oysters that can withstand viruses and an 'enviropig,' whose feces are less harmful to the environment because they contain less phosphorus."[14] The TIMES went on to say that, "...[C]ritics and even some Clinton administration officials say genetically engineered creatures are threatening to slip through a net of federal regulations that has surprisingly large holes.... United States regulators interviewed could not point to any federal laws specifically governing the use or release of genetically engineered animals."

The Clinton/Gore administration announced last week that it will "strengthen" the regulatory system for genetically engineered foods but said the new regulations will definitely not require GE products to carry a label, despite overwhelming public demand for labels. Thus the government's latest regulatory initiative makes one thing crystal clear: what the Clinton/Gore administration and the biotech companies fear most is an informed public.

It will take years before anyone knows what the new regulations entail, or how effective they prove to be. By that time, there may have been hundreds of genetically modified plants and animals introduced into the environment with little or no regulatory oversight. The public is legitimately concerned about this.

In response to these legitimate concerns, the biotech corporations have begun to spend tens of millions of dollars on a public relations campaign because "the public has the right to know more about the benefits of biotechnology." Details next week.

--Peter Montague (National Writers Union, UAW Local 1981/AFL-CIO)

=====

[1] Carey Goldberg, "1,500 March in Boston to Protest Biotech Food," NEW YORK TIMES March 27, 2000, pg. A14.

[2] Marian Burros, "Eating Well; Different Genes, Same Old Label," NEW YORK TIMES September 8, 1999, pg. F5.

[3] Marian Burros, "Eating Well; Chefs Join Effort to Label Engineered Food," NEW YORK TIMES December 9, 1998, pg. F14.

[4] Marian Burros, "U.S. Plans Long-term Studies on Safety of Genetically Altered Foods," NEW YORK TIMES July 14, 1999, pg. A18.

[5] David Barboza, "Biotech Companies Take On Critics of Gene-Altered Food," NEW YORK TIMES November 12, 1999, pg. A1.

[6] Bloomberg News, "New Crop is Said to Aid Nutrition," NEW YORK TIMES December 10, 1999, pg. C20.

[7] "Eating Well; What Labels Don't Tell You (Yet)," NEW YORK TIMES February 9, 2000, pg. F5.

[8] David J. Morrow, "Rise and Fall of 'Life Sciences'; Drugmakers Scramble to Unload Agricultural Units," NEW YORK TIMES January 20, 2000, pg. C1.

[9] Susan Warren, "DuPont Returns to More-Reliable Chemical Business -- Plans for Biotech, Drug Divisions Fizzle as Mergers Change Landscape," WALL STREET JOURNAL February 23, 2000, pg. B4.

[10] Scott Kilman, "McDonald's, Other Fast-Food Chains Pull Monsanto's Bio-Engineered Potato," WALL STREET JOURNAL April 28, 2000, pg. B4.

[11] David Barboza, "In the Heartland, Genetic Promises," NEW YORK TIMES March 17, 2000, pg. C1.

[12] David Barboza, "Monsanto Faces Growing Skepticism On Two Fronts," NEW YORK TIMES August 5, 1999, pg. C1.

[13] National Research Council, GENETICALLY MODIFIED PEST-PROTECTED PLANTS: SCIENCE AND REGULATION (Washington, D.C.: National Academy Press, 2000). ISBN 0309069300. Pre-publication copy available at http://www.nap.edu/html/gmpp/.

[14] Carol Kaesuk Yoon, "Altered Salmon Leading Way to Dinner Plates, But Rules Lag," NEW YORK TIMES May 1, 2000, pg. A1.

Rachel's Environment & Health News is a publication of the Environmental Research Foundation, P.O. Box 5036, Annapolis, MD 21403. Fax (410) 263-8944; E-mail: erf@rachel.org. Back issues available by E-mail; to get instructions, send Email to INFO@rachel.org with the single word HELP in the message. Subscriptions are free. To subscribe, E-mail the words SUBSCRIBE RACHEL-NEWS YOUR FULL NAME to: listserv@lists.rachel.org NOTICE: Environmental Research Foundation provides this electronic version of RACHEL'S ENVIRONMENT & HEALTH NEWS free of charge even though it costs our organization considerable time and money to produce it. We would like to continue to provide this service free. You could help by making a tax-deductible contribution(anything you can afford, whether $5.00 or $500.00). Please send your tax- deductible contribution to: Environmental Research Foundation, P.O. Box 5036, Annapolis, MD 21403-7036. Please do not send credit card information via E-mail. For further information about making tax-deductible contributions to E.R.F. by credit card please phone us toll free at 1-888-2RACHEL. --Peter Montague, Editor

From Montague, P., "#695-Biotech in Trouble — Part 1, May 04, 2000," *Rachel's Environment & Health News, Environmental Research Foundation, P.O. Box 5036, Annapolis, MD 21403. Fax (410) 263-8944; E-mail: erf@rachel.org. With permission.*

Environmental Research Foundation Home

Rachel's Environment & Health News

#696 - Biotech In Trouble--Part 2, May 11, 2000

Biotech In Trouble--Part 2

We saw last week that the genetically-engineered-food industry may be spiraling downward. Last July, U.S. Secretary of Agriculture Dan Glickman -- a big supporter of genetically engineered foods -- began comparing agricultural biotechnology to nuclear power, a severely-wounded industry.[1] (Medical biotechnology is a different industry and a different story because it is intentionally contained whereas agricultural biotech products are intentionally released into the natural environment.)

In Europe, genetically engineered food has to be labeled and few are buying it. As the NEW YORK TIMES reported two months ago, "In Europe, the public sentiment against genetically engineered [GE] food reached a ground swell so great that the cultivation and sale of such food there has all but stopped."[2] The Japanese government also requires GE foods to be labeled. Americans in overwhelming numbers (80% to 90% or more) have indicated they want GE foods labeled but the GE firms consider a label tantamount to a skull and crossbones and the Clinton/Gore administration has sided with the biotech corporations against the people. To be fair, there are no indications that a Republican president would take a different approach. The biotech firms have invested heavily in U.S. elections and the resulting government represents their interests at home just as it does abroad. On this issue, to an astonishing degree, the biotech firms ARE the government.

Since the early 1980s, biotech corporations have been planting their own people inside government agencies, which then created a regulatory structure so lax and permissive that biotech firms have been able to introduce new genetically modified foods into the nation's grocery stores at will. Then these same "regulators" have left government and taken highly-paid jobs with the biotech firms. It represents an extreme case of the "revolving door" syndrome.

The U.S. regulatory system for GE foods, which was created in 1986, is voluntary.[3,pg.143] The U.S. Department of Agriculture regulates genetically engineered plants and the U.S. Food and Drug Administration (FDA) regulates foods made from those plants. If any of the plants are, themselves, pesticidal then U.S. Environmental Protection Agency gets involved. But in no case has any long-term safety testing been done. As the NEW YORK TIMES reported last July, "Mr. Glickman [U.S. Secretary of Agriculture] acknowledged that none of the agencies responsible for the safety of genetically modified foods -- the Agriculture Department, the F.D.A., and the Environmental Protection Agency -- had enough staff or resources to conduct such testing."[1] At the time Mr. Glickman made his statement, 70 million acres in the U.S. had already been planted with genetically modified crops and 2/3rds of the food in U.S. grocery stores contained genetically modified plant materials.[3,pg.33]

The importance of safety testing was emphasized by the National Academy of Sciences (NAS) in its latest (April 2000) report on biotech foods. The NAS [pg. 63] said safety problems might include these:

** New allergens may be introduced into foods.

** New toxins may be introduced into foods. The NAS said, "...there is reason to expect that organisms in US agroecosystems and humans could be exposed to new toxins when they associate with or eat these plants." [pg. 129]

** Existing toxins in foods may reach new levels, or may be moved into edible portions of plants. ("Overall increases in the concentrations of secondary plant chemicals in the total plant might cause toxic chemicals that are normally present only in trace amounts in edible parts to be increased to the point where they pose a toxic hazard," NAS said on pg. 72.)

** New allergens may be introduced into pollen, then spread into the environment. [The NAS remains silent on the human-health implications of new allergens spread via pollen. If the biotech firms have their way, we will learn about this by trial and error. Unfortunately, trial and error has a serious drawback in this instance: once new genetic materials are released into the environment, they cannot be retrieved. Unlike chemical contamination, biotech contamination is irreversible.]

** Previously unknown protein combinations now being produced in plants might have unforseen effects when new genes are introduced into the plants;

** Nutritional content of a plant may be diminished. [pg. 140]

The mechanism for creating unexpected proteins or unexpected toxins or allergens would be pleiotropy, the NAS explained [pg. 134]. Pleiotropy is the creation of multiple effects within an organism by adding a single new gene. In other words, putting a new gene into a tomato, intending to make the tomato more resistant to cold weather, might by chance, and quite unexpectedly, make some people allergic to the new tomato. "Such pleiotropic effects are sometimes difficult to predict," the NAS said. [pg. 134] The NAS said that FDA, USDA and EPA all need to pay attention to such "unintended compositional changes" of genetically modified foods.

Unfortunately, as the NAS pointed out, current tests are not adequate for determining all the problems that might occur because of pleiotropic effects. For example if a new protein is created that has not previously been found in the food supply, then there is no reliable basis for predicting whether it may cause allergic reactions. Allergic reactions are not a trivial matter, the NAS pointed out: "...food allergy is relatively common and can have numerous clinical manifestations, some of which are serious and life-threatening." [pg. 67]

New tests should be developed to test for allergenicity of genetically modified foods, the NAS said several times (see, for example, pg. 8, where the NAS called such new tests "highly desirable"). Specifically, the NAS recommended that tests be developed that actually measure reactions of the human immune system, which is the human system in which allergic reactions develop. The genetically modified foods on the market today have not undergone controlled experiments on real human immune systems. (Putting such foods into grocery stores is an uncontrolled experiment of sorts, but with no one collecting the data.)

In addition to human health problems, the NAS report discussed some of the agricultural and environmental problems that might occur from genetically modified (GM) plants:

** New chemicals in GM plants might kill predators and parasites of insect pests, thus leading to the loss of nature's own biological controls on certain pests. [pg. 74]

** Plants themselves might become toxic to animals. [pg. 75]

** Fallen leaves from GM plants might change the biological composition of the soil, leading to changes in nutrient uptake into plants or even toxicity to creatures living in the soil. [pg. 75]

** Genes from genetically-engineered plants will escape and enter into wild species. This is called gene flow and the NAS says, "[T]otal containment of crop genes is not considered to be feasible when seeds are distributed and grown on a commercial scale." [pg. 92] In other words, gene flow is going to occur. Wild plants are going to receive genes from genetically modified organisms. The biotech firms are re-engineering nature without understanding the means or the ends.

** When a plant is genetically engineered so that the plant itself becomes pesticidal (for example, Bt-containing corn, potatoes and other crops now planted on tens of millions of acres in the U.S.), there may be effects on non-target organisms. In other words, pesticidal crops may affect creatures besides the specific pest they were intended to kill. The NAS says, "Nontarget effects are often unknown or difficult to predict." [pg. 136]

In sum, agricultural biotechnology has raced ahead at lightning speed (going from zero acres planted with GE crops in 1994 to 70 million acres planted in 1999) without any long-term testing, and with minimal understanding of the consequences. The NAS refers to these politely as "uncertainties" and it acknowledges that these uncertainties "often force agencies to base their decisions on minimal data sets." [pg. 139]

So 2/3rds of the food in U.S. grocery stores contains plant materials that were genetically engineered. If they were subjected to government approval at all, it was on a strictly voluntary basis, and the government "often" approved new plants and new foods based on "minimal data sets," according to the National Academy of Sciences. Some of the most important aspects of these new foods had to be ignored because there is no way at present to test for them.

In sum, the biotech industry and its acolytes in government are flying blind and we are all unwitting passengers in their rickety plane. This is not a historical record that inspires confidence. No wonder the Clinton/Gore administration and the biotech corporations do not want anyone to know which foods have been genetically engineered. None of the biotech firms are even CLAIMING that there are taste or nutritional benefits in the biotech foods being sold today, so, to put it bluntly, consumers would have to be out of their minds to eat this stuff or serve it to their children.

Given the serious problems that the NAS said may occur as thousands of new genetically modified foods are introduced into the U.S. food supply without labels, naturally one wonders about liability insurance for the biotech industry. You will not find liability insurance discussed on the biotech industry's web site, www.whybiotech.com, so it is probably one of the industry's most serious problems.

Recently the Swiss company, Swiss Re, issued a report on GE foods.[4] Swiss Re is a re-insurance company -- it insures insurance companies against catastrophic loss. Swiss Re said genetic engineering "represents a particularly exposed long-term risk" and "genetic engineering losses are the kind which have not yet, or only rarely, occurred and whose consequences are extremely difficult to predict."

Swiss Re then asked (and answered) the question, "...so how can genetic engineering risks be insured?" Here is Swiss Re's answer:

"It is currently not possible to give a direct answer to this question. A lot depends on whether consensus can be reached on the relevant loss scenarios in a dialogue involving the genetic engineering industry, society, and the insurance industry. This will make genetic engineering risks more calculable and more interesting to traditional insurance models. From the point of view of the insurance industry, WE ARE AT PRESENT A LONG WAY OFF. [Emphasis added.]

"Today we must assume that the one-sided acceptance of incalculable risks means that any participants in this insurance market run the risk not only of suffering heavy losses, but also of losing control over their exposure."

Without intending to do so, the Swiss Re report brings to mind an agenda for citizens who oppose the expansion of ag biotech:

(a) On the principle that the polluter shall pay, biotech firms should be held strictly liable for any harms they may cause, not requiring proof of negligence;

(b) Ag biotech corporations should not be allowed to self-insure; as we know from the asbestos industry, self-insurance can lead to bankruptcy and hundreds of thousands of legitimate claims never being paid;

(c) Law suits should seek damages for gene flow, pollen drift, inadequate testing for allergenicity, crop failures, and so on. A series of lawsuits against private firms or government agencies would get the insurance industry's attention.

(d) Stockholders in ag biotech firms should express concern (to the board of directors, and to the Securities and Exchange Commission) about the failure to disclose incalculable risks. Stockholders in insurance companies should express concern about the potential for "heavy losses" and "losing control over their exposure" if coverage is extended to ag biotech firms.

--Peter Montague (National Writers Union, UAW Local 1981/AFL-CIO)

=====

[1] Marian Burros, "U.S. Plans Long-Term Studies on Safety of Genetically Altered Foods," NEW YORK TIMES July 14, 1999, pg. A18.

[2] Carey Goldberg, "1,500 March in Boston to Protest Biotech Food," NEW YORK TIMES March 27, 2000, pg. A14.

[3] National Research Council, GENETICALLY MODIFIED PEST-PROTECTED PLANTS: SCIENCE AND REGULATION (Washington, D.C.: National Academy Press, 2000). ISBN 0309069300. Pre-publication copy available at http://www.nap.edu/html/gmpp/.

[4] Swiss Re, GENETIC ENGINEERING AND LIABILITY INSURANCE; THE POWER OF PUBLIC PERCEPTION (UNDATED). Available from http://www.swissre.com/e/publications/publications/flyers1/- genetic.html (omit the hyphen).

Rachel's Environment & Health News is a publication of the Environmental Research Foundation, P.O. Box 5036, Annapolis, MD 21403. Fax (410) 263-8944; E-mail: erf@rachel.org. Back issues available by E-mail; to get instructions, send Email to INFO@rachel.org with the single word HELP in the message. Subscriptions are free. To subscribe, E-mail the words SUBSCRIBE RACHEL-NEWS YOUR FULL NAME to: listserv@lists.rachel.org NOTICE: Environmental Research Foundation provides this electronic version of RACHEL'S ENVIRONMENT & HEALTH NEWS free of charge even though it costs our organization considerable time and money to produce it. We would like to continue to provide this service free. You could help by making a tax-deductible contribution(anything you can afford, whether $5.00 or $500.00). Please send your tax- deductible contribution to: Environmental Research Foundation, P.O. Box 5036, Annapolis, MD 21403-7036. Please do not send credit card information via E-mail. For further information about making tax-deductible contributions to E.R.F. by credit card please phone us toll free at 1-888-2RACHEL. --Peter Montague, Editor

From Montague, P., "#696-Biotech in Trouble — Part 2, May 11, 2000," *Rachel's Environment & Health News, Environmental Research Foundation, P.O. Box 5036, Annapolis, MD 21403. Fax (410) 263-8944; E-mail: erf@rachel.org. With permission.*

---

http://www.rachel.org/bulletin/bulletin.cfm?Issue_ID=1763

Invading aliens are the stuff of science fiction. But Monsanto's plan to genetically engineer food is all too real, turning U.S. crops into ...

The X Fields

Monsanto's genetically engineered soybeans have not been sufficiently tested and the risks to human health are unknown.

The camera zooms in on peculiar green stalks, then pulls back to reveal endless rows of a crop, immaculate lines running uninterrupted for miles — except for an immense and inexplicable "X" spanning the field. A "crop circle" of sorts. Not evidence of a UFO visit but certainly of unnatural activity.

But the alien presence is the crop itself. It is genetically mutated soy beans, a different form of alien invasion. And it comes not from outer space, but from the U.S. chemical giant Monsanto which has developed and is selling the first genetically engineered soy beans.

Monsanto's so-called "Roundup Ready" soybean, has been genetically engineered through the use of genes from viruses and bacteria to be resistant to Monsanto's own "Roundup" herbicide. The "Roundup Ready" soybeans encourage farmers to apply the herbicide two to three times or more during the plant's life cycle, allegedly without harming the soybean crop. If "Roundup" were sprayed directly on a normal soybean plant, it would kill it.

Dosing countless acres with this toxic herbicide will convert the land to a dead zone, barren of all life except the mutant soybeans.

The genetically engineered (G-E) beans are not cheaper, tastier or healthier. Nor do they produce a higher yield. There is absolutely no benefit in this plot for the consumer, so why is Monsanto doing this? The answer, as is all too often the case, is profit.

Monsanto is pitching its G-E bean as a total weed management system. The farmers buy the beans and the herbicide together. This secures the market for Monsanto's products after its "Roundup" patent expires in 2000.

INTO THE UNKNOWN

The risks to human health and the environment posed by G-E food are unknown because this is a completely new type of food.

One significant risk associated with G-E food is allergy. A previous attempt to genetically engineer soy using a gene from the brazil nut was abandoned because it was discovered that people allergic to brazil nuts were also allergic to the G-E bean. Brazil nuts are a known allergen so it was easy to test for this, but some genetically engineered products are using genes which have never been part of the human diet.

"By introducing these soybeans into the global marketplace, Monsanto is treating consumers like guinea pigs for their genetic experiments," said Greenpeace's Genetic Engineering campaigner Beth Fitzgerald.

SUPER UN-NATURAL

Once introduced, G-E soy beans could wreak havoc on nature. These new types of plants could escape into the wider environment and displace existing plants. Once introduced, new species can prompt irreversible changes with disastrous results, as humans have already learned the hard way.

Genetically engineered plants and animals are completely new life forms. With no natural habitats outside the laboratory, they are untested in the wider environment. G-E plants can breed and even cross-breed, possibly endangering whole ecosystems. Monsanto's plan is a lunatic experiment with the web of life.

THE TRUTH IS OUT THERE

Because manipulating nature can have serious repercussions, many of them unforeseen and unforeseeable, Greenpeace stresses caution for the environment and health. Lacking assurance that this experiment is safe, the most judicious course of action must be taken. It is, after all, better to be safe than be sorry. The responsibility for proving the safety of the G-E soybean lies with Monsanto.

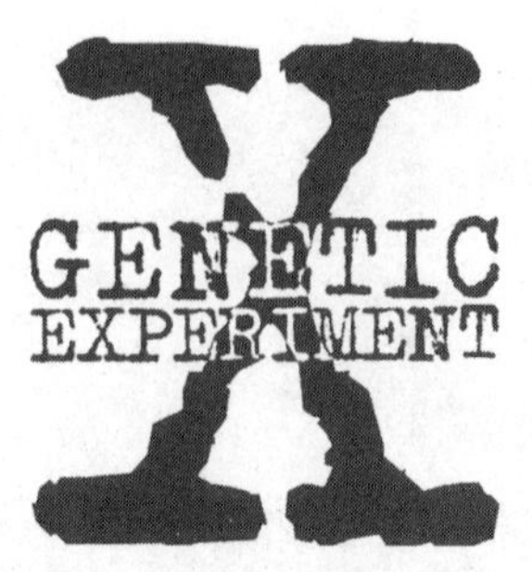

The G-E soybeans are not the first questionable contribution Monsanto has made to society. Monsanto developed Agent Orange, the controversial toxic defoliant used in Vietnam, and was one of the original manufacturers of PCBs. The chemical corporation introduced rBGH, a bovine growth hormone, which caused serious health problems in cows. Earlier this year, Monsanto's genetically engineered cotton, designed to resist a cotton pest, failed.

One of the world's most profitable multinational companies, Monsanto is investing billions of dollars in genetic engineering, an estimated $2 billion since the late 1970s.

Greenpeace activists "quarantined" this field of genetically engineered soybeans before harvest.

© SCOTT OLSEN/GREENPEACE 1996

SAY IT AIN'T SOY

Monsanto claims that G-E soybeans are safe and that they tested them (yes, but for a mere 10 weeks). The company hasn't evaluated the risk of gene transfer or tested the G-E bean to establish its potential to provoke allergies. The bigger test is being conducted in the wider environment and on you.

The first G-E crops were harvested in the U.S. in September and October, then mixed with normal soybeans, and shipped to Europe. Although only 1 to 2 percent of all soy coming from the U.S., these will be mixed with the G-E-free beans, thus contaminating the entire supply. Half of the U.S. soybean crop is for the domestic market, and half is exported, mostly to Europe.

In recent months, some of Europe's largest food producers and wholesalers—Unilever and Nestlé in Germany, Safeway in the United Kingdom, for example—have bowed to pressure from Greenpeace and stated that they will not use genetically engineered soybeans. Previously, they had said they must use the G-E soybeans because the beans would already be mixed. In fact, not only is separation possible, but a number of suppliers and brokers have already started doing it in anticipation of great demand for normal, non-G-E soybeans.

Monsanto will reap all the profits and leave all the risk to the family farmers. If European markets reject the G-E soybeans, 25 percent or more of the U.S. market could vanish leaving farmers with unsold and unwanted crops and poisoned land.

CURDS AND —WHOA!

"When I heard about Monsanto's plans for genetically engineered soy beans, I was ready to put on my armor and go to battle," said Luke Lukoskie. He is president and owner of Island Spring, Inc., a producer of organic soy-based foods, on Vashon Island in Washington state's Puget Sound region. For 20 years, Lukoskie has been committed to using naturally occurring, organic soy beans because of his concern for the health of consumers and for the health of the land.

© BETH FITZGERALD/GREENPEACE 1996

In Iowa, Greenpeace protesters left this "warning sign" in a field of genetically engineered soybeans. Monsanto's own sign (left) has no warning.

"There are 10,000 natural varieties of soybeans — plenty to choose from — we simply don't need a genetic mutation of an already exceptional selection. We are totally opposed to the genetic engineering of soybeans and the increased use of chemicals on the land. Monsanto's plan is definitely not a healthy thing to have happen to the planet," Lukoskie said.

EXTRA-TERRESTRIAL ENCOUNTER

Early in the morning on the day before the harvest, 30 Greenpeace activists "quarantined" a field of genetically engineered soybeans owned by Monsanto and blocked the harvest.

Borrowing a theme from television's *The X Files*, our activists marked Monsanto's field with a huge 100-foot "X," painted with bright pink, non-toxic, milk-based paint. The field was not harmed in the protest.

"Strange things are happening to America's crop fields—they're turning into America's 'X fields,'" said Fitzgerald. "Monsanto is introducing alien species into the environment at a potentially high cost to humans, the environment and the fiscal future of U.S. farmers."

THE COMING CONSUMER CLAMOR

The concern over G-E soybeans is global. Greenpeace activists in Germany, Sweden, Austria, and England have held protests in recent months against Unilever, Europe's largest food company. In October, after being the target of two Greenpeace actions, Unilever Germany said that it would not be using G-E soy in margarine and vegetable oils and that it would push the soy industry to separate conventional soybeans from the G-E soybeans. As we went to press, other European companies were announcing they would exclude Monsanto's G-E soybeans from their product lines.

Consumers have a right to G-E-free food. Genetically manipulated soybeans have no benefits whatsoever for the consumer and could harm the web of life.

If this first harvest of G-E soybeans makes it into processed foods, each product should be labeled as containing G-E soy, so that consumers can make informed decisions about what they eat. No one should be made part of Monsanto's experiment unwittingly.

For more information, please read the Greenpeace report "Not Ready for Roundup," which is available on our website at: **http://www.greenpeace.org/~usa /reports/biodiversity/roundup** ●

From *Greenpeace Magazine,* 1996, "The X Fields." With permission.

Genetically Engineered Foods: Who's Minding the Store

Q What are genetically engineered foods?

A The edible portions of genetically engineered plants or animals (e.g. tomatoes from genetically engineered tomato plants) are what most people mean when they speak of genetically engineered foods. Genetically engineered plants and animals are modified by modern genetic techniques, such as recombinant DNA, which allow researchers to modify genetic material in ways not possible with traditional selective breeding. For example, researchers can transfer genetic material from one species to another, such as from animals to plants.

In some cases chemical additives manufactured by genetically engineered bacteria are also called genetically engineered foods. For example rennin, an enzyme used in cheese manufacturing, is extracted from bacteria engineered with a copy of a cow gene. However, unlike consumers of genetically engineered tomatoes, consumers of engineered additives do not directly consume genetically engineered organisms.

Q What's coming to grocery store shelves?

A A wide variety of genetically engineered crop plants are now under development, and some crops have reached the marketplace. As of August 1995, the U.S. Department of Agriculture had reviewed more than 1500 submissions for field trials of genetically engineered crops. Crops now in commercial production include tomatoes altered with a synthetic gene that retards softening; potatoes and corn with bacterial genes for insecticidal toxins; soybeans and cotton (some grown for cottonseed oil) with bacterial genes that allow the crops to tolerate applications of chemical weedkillers; and squash with viral genes that confer disease resistance. Although their development is not as far along, livestock and fish are also being genetically engineered.

Q Are genetically engineered foods dangerous?

A Although most are likely to be safe, some may not be. To consumers, most genetically engineered foods are essentially foods with added substances -- usually proteins. This is because genes are "translated" into proteins by cells. Therefore, when a genetic engineer adds, say, a bacterial gene to a tomato, he or she is essentially adding a bacterial protein to that tomato. In most cases these added proteins will likely prove safe for human consumption. Nevertheless, just as with conventional food additives, substances added to foods via genetic engineering may in some instances prove hazardous.

A major concern about adding proteins to foods via genetic engineering is that they may cause susceptible individuals to become allergic to foods they previously could safely consume. Food allergies are a serious public health concern, which food allergists estimate affect roughly 2.5 - 5 million Americans. Allergic reactions cause discomforts and in some cases life-threatening anaphylactic shock. Since virtually all known food allergens are proteins, foods with new proteins added via genetic engineering could sometimes become newly allergenic. These concerns about food allergy are real. One company has already dropped plans to commercialize soybeans with a brazil nut gene after testing revealed the soybeans were likely to cause allergic reactions in brazil nut allergic individuals. Unfortunately, food allergies are poorly understood, and in many cases scientists will not be able to test the potential allergenicity of genetically engineered foods.

Q Does the FDA's policy for foods from genetically engineered crops safeguard consumers?

A FDA's policy, announced by former Vice President Dan Quayle in May, 1992, as "regulatory relief," appears to do more to protect the biotechnology industry than to protect consumers. FDA's policy includes a series of "decisions trees" for industry decision-making, a series of yes-no questions, such as "Is there any reported toxicity? or "Does the biological function raise any safety concern?" Food producers are then supposed to decide for themelves whether they need to consult FDA before they market foods obtained from genetically engineered crops.

FDA, at least in principle, is applying the same regulations to substances added to foods via genetic engineering as apply to conventional chemicals added to food. But, FDA's decision trees appear to significantly weaken a longstanding requirement under food safety law: Food manufacturers must establish scientifically the safety of new substances added to food before selling them to the public, regardless of whether the manufacturers think they are safe. FDA's policy states that the agency will only require approval "in cases where safety questions exist sufficient to warrant formal premarket review". Deciding if such questions exist is left to food manufacturers.

Q What about food labeling?

A Under the 1992 policy, FDA will only require labeling of genetically engineered foods under certain exceptional circumstances. Since most genetically engineered foods will be indistinguishable in appearance from nonengineered foods, consumers will generally not know what they are buying. FDA ignores consumers' right to know by ignoring longstanding regulations that require in most circumstances that manufacturers label foods to disclose their ingredients. For example, researchers have genetically engineered vegetables to produce a new protein sweetener. Existing FDA regulations mandate that companies disclose sweeteners added to canned vegetables via conventional means. Yet, FDA will not require that protein sweeteners added to vegetables via genetic engineering be labeled as ingredients.

Labeling is vital to food allergic individuals, who need to now when their purchases are potentially allergenic. FDA will require labeling of foods genetically engineered to contain potential allergens from only the most commonly allergenic foods -- a requirement that threatens individuals with less common food allergies.

Some vegetarians and individuals who follow religious dietary laws have told FDA that they want to know when animal genes are added to plants used as foods. FDA has taken no steps to accommodate their dietary beliefs and restrictions.

http://www.environmentaldefense.org/pubs/FactSheets/a_BioTFact.html

Portrait Of An Industry In Trouble

by Brian Halweil

Real Audio clips from the February 17, 2000 Biotechnology Forum:
Clips 1-11: Remarks by the full panel including a 2 Q&A clips (length: over 1 hour)

After four years of stupendous growth, farmers are expected to reduce their planting of genetically engineered seeds by as much as 25 percent in 2000, as spreading public resistance staggers the once high-flying biotech industry. (See Figure 1.) Stock prices for agricultural biotech companies are falling, exports of transgenic crops are tumbling, and questions are mounting about the liability for what is turning into a major debacle for farmers. At the same time, some 130 nations just signed an international biosafety agreement prescribing caution.

Worldwide, the area planted to transgenic crops jumped more than twenty-fold in the last four seasons, from 2 million hectares in 1996 to nearly 40 million hectares in 1999. In the United States, Argentina, and Canada, over half the acreage for major commodities like soybeans, corn, and canola are planted in transgenics. (These three nations account for 99 percent of the global transgenic acreage, pointing to the limited global acceptance.)

Figure 1: Global Area of Transgenic Crops, 1996-99, with Projections for 2000

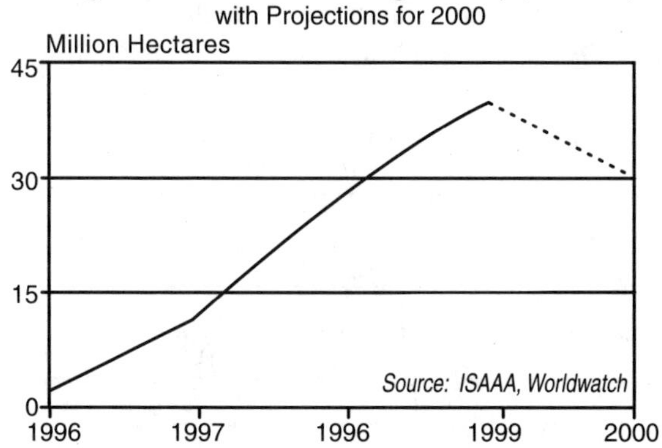

But with a growing number of food manufacturers and grocery chains in Europe taking products containing transgenics off the shelves, the market for these crops has been shrinking. American exports of soybeans to the European Union plummeted from 11 million tons in 1998 to 6 million tons last year, while American corn shipped to Europe dropped from 2 million tons in 1998 to 137,000 tons last year: a combined loss of nearly one billion dollars in sales for American agriculture.

Investors have reacted harshly to the growing consumer rejection of transgenics and the resulting reduced sales of engineered seed and complementary agrochemicals. In

May of 1999, Europe's largest bank, Deutsche Bank, recommended that investors sell all holdings in companies involved in genetic engineering, declaring that "GMO's [Genetically Modified Organisms] Are Dead." The bank's report envisioned the development of a two-tiered commodity market in which non-transgenic crops would command price premiums over transgenic crops—a prospect that threatens the farmers planting engineered seeds and the companies that sell these seeds.

In fact, top commodity handlers, such as Archer Daniels Midland and A.E. Staley, have already begun to discount transgenic crops because of this greater financial risk. Commodity traders have followed suit fearing the loss of export markets as Japan, South Korea, Australia, Mexico, the members of the European Union, and other nations draft laws requiring mandatory labeling of food products containing transgenic ingredients.

Most major food companies have already announced that they will avoid transgenic ingredients in their products for the European market. But now recent surveys indicate that consumer tastes are souring on the other side of the Atlantic as well. Several food manufacturers, including Gerber, Frito-Lay, and natural food retailers Wild Oats and Whole Foods, have said that they will avoid transgenic ingredients in their products sold in the United States—the largest consumer market for transgenic crops. If more American manufacturers hop on the bandwagon, the drop in demand would be devastating for transgenic growers and seed producers.

Share prices for biotech seed companies that were Wall Street's darlings a few years ago are sinking towards all-time lows. Investors in Monsanto Company, the industry leader which has born the brunt of public criticism, have watched the corporation's share price lose nearly one-third of its value in the last year, falling from a high of $50 in February of 1999 to a recent low of just $35. (See Figure 2.)

Figure 2: Monsanto Company Share Price, January 1997 to February 2000

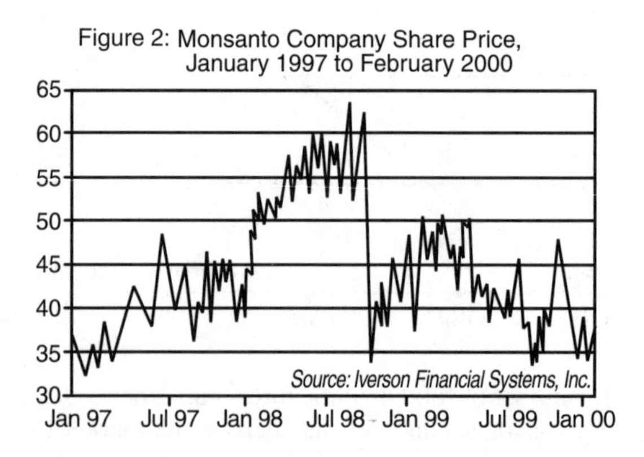

Brokerage houses have been advising major players in the biotech industry to spin off their ailing agricultural divisions. Novartis and AstraZeneca both followed this advice in December of 1999. Dupont had been considering issuing a new stock that would track its much-celebrated and nascent ag biotech division, but decided in early 2000 to indefinitely postpone the stock's release. And struggling to recoup nearly $8 billion in seed company and agricultural biotechnology investments, Monsanto merged with pharmaceutical and chemical giant Pharmacia Upjohn at the end of 1999. The new firm quickly decided to turn Monsanto's agricultural unit into a separate company

Further complicating the financial picture are concerns about uninsured liabilities for farmers and agribusiness companies. In November 1999, 30 farm groups, including the National Family Farm Coalition and the American Corn Growers Association, warned American farmers that "inadequate testing of gene-altered seeds could make farmers vulnerable to 'massive liability'from damage caused by genetic drift—the spreading of biologically modified pollens–and other environmental effects." In December, a group of high-profile lawyers filed a class-action lawsuit against Monsanto, on behalf of American soy farmers, charging that the company has not conducted adequate safety testing of engineered crops prior to release and that the company has tried to monopolize the American seed industry.

To many observers, the rapid release of transgenic crops and the ensuing financial disarray is disturbingly reminiscent of the earlier uncritical bandwagons for nuclear energy and chemical pollutants like DDT. A combination of public opposition and financial liability eventually forced retrenchment of these earlier technologies, after their effects on the environment and human health proved to be far more complex, diffuse, and lingering than the promises that accompanied their rapid commercialization.

In an effort to avoid this same dismal cycle with the introduction of each new "revolutionary" technology, public policy advocates have called for the adoption of the precautionary principle. Under current policy, a technology is all too often judged safe until it is definitively proven harmful. The precautionary principle holds that when a new technology carries suspected harm, scientific uncertainty of the scope and scale of the harm should not necessarily prevent precautionary action. Instead of requiring critics to prove that the technology poses potential dangers, the producers of a technology shoulder the burden of presenting evidence that the technology is safe.

Industry has long labeled the precautionary approach as reactionary, arguing that it stifles research and prevents economic progress. On the contrary, advocates realize that all stakeholders—including consumers, government, and industry—benefit from an open and democratic attempt to anticipate any undesirable social and financial surprises. The goal is to apply wisdom and judgement about the potential effects of a new technology before flooding the marketplace with the products of that technology.

The rapid rollout of genetically engineered crops over the last four years stands the precautionary principle on its head. Widespread commercialization of transgenic crops has come before—not after—any thorough examination of the benefits and risks associated with these crops. The regulatory framework devoted to transgenics is inadequate, nontransparent, or completely absent. And there has been essentially no public discussion about the many potential consequences of large-scale planting of transgenic crops. For example, U.S. Secretary of Agriculture Dan Glickman only recently called for studies assessing the long-term ecological effects of these crops. But more than half of the U.S. soybean crop and nearly as much of the corn crop are already genetically engineered.

Another recent illustration of our lack of precaution was presented in a December 1999 article in *Nature* reporting that the insecticide produced by a widely planted variety of transgenic corn can accumulate—in its active form—in the soil for extended periods of time. The authors note that the effects on soil organisms and soil fertility are largely unknown, but potentially enormous. But, like earlier laboratory studies showing that pollen from this same corn could be lethal to certain beneficial insects, the fact that such effects had not been considered prior to planting tens of millions of hectares in this crop raises concerns about the adequacy of existing safeguards for ecological and human health risks.

ALSO SEE:
Audio clips from a group of panelists who partook in the Agriculture biotech brief on February 17, 2000.

FOR MORE INFORMATION CONTACT:

Worldwatch Institute
1776 Massachusetts Ave NW
Washington, DC 20036
telephone: 202 452-1999
fax: 202 296-7365

The next endangered species?
Take Action Packet
The Campaign
To Label Genetically Engineered Foods

Congressional labeling legislation would protect consumers' right to know

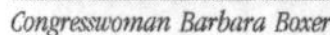

Congresswoman Barbara Boxer

Congressman Dennis Kucinich

If you're like most Americans, chances are you'd like genetically engineered foods to be labeled. Surveys routinely indicate, after all, that a large majority of Americans want labeling.

So far, giant agribusinesses and the U.S. government have balked at the idea of putting labels on genetically engineered foods. But two bills introduced in Congress in the past few months offer hope that the wishes of the American people may soon be granted—especially with your help.

House Resolution 3377 and Senate Bill 2080, both named the Genetically Engineered Food Right to Know Act, would require any genetically engineered foods, or any foods that contain genetically engineered ingredients, to bear the following label:

"This product contains genetically engineered material, or was produced with a genetically engineered material."

Americans are demanding labeling for a number of reasons. Some people have allergies, and worry that genes taken from a food or substance to which they are allergic may be placed in other foods they might buy, without their knowledge. Some vegetarians are concerned that genes from animals, or even humans, may be inserted into fruits or vegetables.

And many people simply want the right to choose whether they will support the genetic engineering industry or not. Several biotech companies also manufacture pesticides and are involved in other forms of nonsustainable agriculture that some people find objectionable.

If you support labeling, be sure to make your opinions known by sending letters to Congress and the presidential candidates. In this action packet, we've provided five clip-and-mail letters that you can send to your two Senators, one Representative, and both Presidential candidates. You also can photocopy them and pass them out to your friends.

Congressman Dennis Kucinich (D-Ohio) introduced HR 3377 late last year. He says that labeling is warranted because genetically engineered foods are widespread, but scientific knowledge about them so far is limited.

An estimated two-thirds of foods on supermarket shelves are genetically modified, or contain genetically modified organisms (GMOs). In 1999, more than one-fourth of American crops were genetically engineered, including 35 percent of all corn and 55 percent of all soybeans.

"It's antiquarian not to have labeling on these foods," Kucinich says. "Industry has taken a very paternalistic attitude toward people."

"There's something very American about it; people want the right to know," he adds. "We're the country of freedom of information."

Genetically engineered foods already are required to be labeled in Great Britain, France, Germany, the Netherlands, Belgium, Luxembourg, Denmark, Sweden, Finland, Ireland, Spain, Austria, Italy, Portugal, Greece, New Zealand and Japan.

Senator Barbara Boxer (D-California) introduced Senate Bill 2080 in February. She supports labeling, she says, because "we don't know whether genetically engineered food is harmful or whether it is safe. Scientists have raised concerns about genetically engineered food. These concerns include the risks of increased exposure to allergens, decreased nutritional value, increased toxicity and increased antibiotic resistance."

She adds that "scientists have raised concerns about the ecological risks associated with genetically engineered food. Some of those risks include the destruction of species, cross pollination that breeds new weeds that are resistant to herbicides, and increases in pesticide use over the long-term."

Both bills would require crops such as genetically engineered potatoes, soybeans and corn to be labeled. They would also mandate the labeling of processed foods that contain GMOs such as soy and corn byproducts. Food containing milk from cows treated with genetically engineered hormones also would require labeling.

Kucinich says his bill would set up an efficient and highly effective system of guarantees to ensure that producers, manufacturers and retailers label all genetically engineered foods. At each stage of production—seed company, farmer, manufacturer, retailer—the business that has custody over the genetically altered food would be required to label it.

If the foods do not contain GMOs, businesses would guarantee that they are GMO-free. That way, the business at the next stage of production would not be subject to civil or criminal penalties if it failed to label a food later found to contain GMOs. Manufacturers and retailers at later stages of the process would be able to rely on guarantees and would not have to repeat testing.

Drugs would not have to be labeled under the bill. Restaurants would not be required to label foods.

The Campaign
to Label Genetically Engineered Foods

Take Action Letters

This *Take Action Packet* contains letters to mail to your House Representative, your two Senators and the Presidential Candidates.

Visit us online

Be sure to visit The Campaign's web site for news updates and educational information, and to purchase books about genetically engineered foods. You will find form letters, our expanded educational tutorial and links to dozens of other web sites. You can even become a member of The Campaign online by signing up on our secured server.

See details online at:
www.thecampaign.org

The Campaign's educational tutorial on genetically engineered foods

Genetically engineered food can be a complicated topic. Luckily for you, we've taken all the hard work out of it! Just sit back and let us guide you through this strange new biotech world. Please share this information with your friends and family. You'll find an extended version of this tutorial online at www.thecampaign.org/brochureindex.htm.

Table of Contents

The simple ABCs of genetic engineering

Some biology basics

Plants and animals are made up of millions of cells. Each cell has a nucleus, and inside every nucleus are strings of DNA (deoxyribonucleic acid, if you want to get technical). DNA contains complete information regarding the function and structure of organisms ranging from plants and animals to bacterium.

A gene represents the blueprint of an animal or plant. Genes determine an organism's growth, size and other characteristics. Genes are made up of sequences of DNA. As you remember from basic biology, genes are the units by which species transfer inheritable characteristics from one generation to the next.

Genetic engineering is the process of artificially tampering with these blueprints. Through genetic engineering, scientists insert the gene of one organism into another in an effort to replicate characteristics in the receiving organism.

So, for example, genetic engineers have injected tomatoes with the antifreeze gene of a flounder in an effort to give the tomato a longer growing season. Genetic engineers also plan to use the technology to improve nutrition and even plan to develop medical benefits. Some biotechnology companies also are developing crops that can withstand increased amounts of pesticides, often pesticides sold by those very same companies.

"We are living today in a very delicate time, one that is reminiscent of the birth of the nuclear era, when mankind stood at the threshold of a new technology," says Dr. John Fagan, a molecular biologist and former genetic engineer. "No one knew that nuclear power would bring us to the brink of annihilation or fill our planet with highly toxic radioactive waste. We were so excited by the power of a new discovery that we leapt ahead blindly, and without caution. Today the situation with genetic engineering is perhaps even more grave because this technology acts on the very blueprint of life itself."

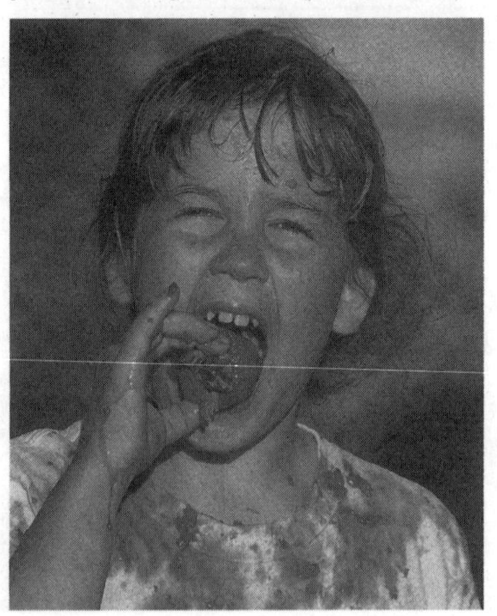

Biotech corporations: Big promises, but can they deliver?

Biotech corporations make bold claims about the ability of genetically engineered foods to change the world—promises ranging from feeding the world's hungry to saving the environment. Here's a look at some industry promises, and some facts that put these assurances in dispute.

Promises & Realities

Promise: Biotech will feed the world's poor.
Reality: Biotech companies are much more interested in the corporate bottom line than in helping the poor.

Consider the case of the "terminator seed" technology, pursued by Monsanto, one of the largest biotech companies. The terminator seed is a genetic engineering technology that sterilizes seeds produced by crops. The technology would force farmers to purchase seeds every year from companies who sell the seeds.

Analysts worry that under terminator technology, many staples for the world's poorest people, including wheat, rice and soybeans, would be under the control of international agribusinesses. Up to 1.4 billion farming families worldwide may be forced to buy into the terminator technology.

Monsanto recently announced that, because of public opposition, it would not commercialize the terminator. However, the company said it will continue to pursue several related gene technologies, and could change its mind about the terminator in the future.

If the multinationals really want to help feed the poor, would they come up with technologies so pernicious?

Promise: Biotech will save the environment.
Reality: Biotech is a risky experiment that may have vast environmental repercussions.

The companies behind genetic engineering don't have a great environmental track record. Some of these companies were behind the development of risky chemicals such as DDT and Agent Orange. As we've learned over the past few decades, the development of deadly pesticides has had disastrous implications for wildlife and human development.

U.S. farmers already have planted millions of acres of GE corn. Three years after GE corn was launched on a massive commercial scale, Cornell University scientists discovered that the mutated corn may be deadly to Monarch butterflies. What other surprises may be in store?

Promise: GE crops require fewer pesticides.
Reality: Biotech companies are using GE technologies to sell higher quantities of the pesticides they manufacture.

Many of the companies behind biotech, in fact, such as Monsanto, DuPont and Novartis, also manufacture toxic pesticides. One of the most popular categories of GE foods are crops that are resistant to pesticides, meaning that more pesticides can be applied. Monsanto, for example, has created the Roundup Ready soybean, which is engineered to withstand higher doses of Monsanto's Roundup pesticide.

Pesticidal potatoes, terminator seeds and genetically mutated trees, oh my!

The world of biotechnology is filled with harrowing tales of strange, new "Frankenfoods." If it's bizarre, genetic engineers can create it! Here are just a few of their frightening creations:

1. Pesticidal potatoes

For years, many chemical companies made a lot of money by selling pesticides to spray on crops. These days, the game is changing: Genetic engineers have created potatoes that actually can produce **their own** pesticides.

The New Leaf Superior, marketed by the Monsanto corporation since 1995, is engineered to produce the insecticide Bt, or Bacillus thuringiensis, in each one of its cells. Bt kills the Colorado potato beetle, one of the biggest threats to healthy potatoes. Unfortunately, the pesticidal potatoes are not labeled, so unless you consume only organic potatoes, there's no way to be sure that you're not eating the pesticidal variety. And some scientists say that the long-term effects of eating these potatoes is unknown.

In 1998, the *New York Times* reported that regulation of the pesticidal potato has fallen through the cracks of the U.S. government. The Food and Drug Administration told the *Times* it does not regulate the potato because it does not have the authority to regulate pesticides; that responsibility, said the FDA, lies with the Environmental Protection Agency. But the EPA said labeling pesticidal potatoes is FDA's job, because potatoes are a food. The FDA responded that the Food, Drug and Cosmetic Act forbids the food agency from including information about pesticides on foods. And so it goes.

Meanwhile, Phil Angell, Monsanto's director of corporate communications, told the *Times* that "Monsanto should not have to vouchsafe the safety of biotech food. Our interest is selling as much of it as possible. Assuring its safety is the FDA's job."

2. Terminator seeds

Monsanto also has developed a new seed technology that has many environmentalists and Third World leaders crying foul.

The "terminator seed," still in developmental stages, is designed to produce sterile crops that don't produce seeds. Under the new technology, Monsanto would force farmers to buy seeds from the giant agribusiness every year. Historically, farmers have saved some seeds from one growing season for use in the next.

Environmentalists worry what will happen when the terminator is unleashed on the environment. They fear that terminator technology could migrate from one farm to another, or from a farm to wild plants. And activists in developing nations, who say that up to 1.4 billion farming families worldwide may be forced to buy the seed, say the technology would put too much power in the hands of a few international agribusinesses.

Responding to the intense criticisms, Monsanto announced in 1999 that it would not commercialize the terminator seed. However, the company continues to research several related technologies, and could change its mind about the terminator down the road.

3. Genetically mutated trees

Genetic engineering is a field that extends into many areas beyond food. One of the more frightening possibilities to arise so far is the genetically mutated forest.

Scientists say that plans for "terminator" trees —engineered never to flower—could create a "silent spring" in the forests. While these trees would grow faster than traditional trees, they would be lifeless by comparison. Gone would be the bees, butterflies, moths, birds and squirrels that depend on pollen, seed and nectar of normally reproducing trees.

"If you replace vast tracts of natural forest with flowerless trees, there will be a serious effect on the richness and abundance of insects," says George McGavin, curator of entomology at Oxford University Museum. "If you put insect resistance in the leaves as well you will end up with nothing but booklice and earwigs. We are talking about vast tracts of land covered with plants that do not support animal life as a sterile means of culturing wood tissue. That is a pretty unattractive vision of the future and one I want no part of."

4. Glow-in-the-dark potatoes

Edinburgh scientists have mixed jellyfish genes with potatoes, resulting in spuds that glow when they need watering. The potatoes are not intended for consumption; only a few would be planted per hectare for water monitoring purposes. But ecologists wonder what would happen if the potatoes got mixed in with the regular batch.

Meteoric growth: GE foods now are almost everywhere you look

If you want to avoid eating genetically engineered foods, all we can say is *good luck*. In just a few short years, GE foods have swept into the marketplace, affecting almost all of the foods we eat. In fact, the only way you can be sure to avoid eating genetically mutated foods is to buy organic, or to grow your own.

The basic facts:

The first large-scale commercial harvest of genetically engineered crops in the United States was in 1996. **By 1999, more than one-fourth of American crops were genetically engineered**, including:

- 35 percent of all corn
- 55 percent of all soybeans
- nearly half of all cotton.

So far, at least **50 GE crops have been approved** by the USDA, including potatoes, tomatoes, melons and beets. GE rice, wheat, cucumbers, strawberries, apples, sugarcane and walnuts are being grown on test sites.

Some experts estimate that **GE ingredients can be found in as many as two-thirds of all items on supermarket shelves.** Even if you shop at the local health food store, you may be eating some genetically engineered foods.

Some common foods that frequently contain GE ingredients:

- tortilla chips
- drink mixes
- taco shells
- veggie burgers
- muffin mix
- baby formulas

Watch out for any foods that contain soybean or corn derivatives. Soy finds its way into about 60 percent of processed foods. GE ingredients include soy oil, soy flour, lecithin, and soy protein isolates and concentrates. Corn products commonly found in processed foods include corn oil, corn starch, corn flour and corn syrup.

Animal products are another high-risk category. Genetically modified organisms may be present in meat, poultry, seafood, milk, cheese, yogurt and whey. Most of the corn and soybeans grown in the United States are fed to farm animals. Also, dairy products may come from cows that have been treated with bovine growth hormone (BGH).

Allergic reactions and other possible health risks

By now, millions of acres of genetically engineered crops have been planted, and nearly two-thirds of the products on our supermarket shelves contain GE ingredients. But GE foods remain poorly studied; scientists simply can't say with any authority that they are absolutely safe for human consumption. In fact, many questions persist.

Essentially, we've been subjected to a massive experiment on human health. What will the results of this experiment be? Stay tuned.

1. Very few studies have been conducted to determine whether genetically engineered foods are harmful to human health.

Genetic engineering is a young, and in many ways poorly understood, technology. Many scientists believe that genetically engineered foods have been rushed much too quickly to market — to boost multinationals' profit margins — before adequate testing has been completed to ensure public health.

According to the *Washington Post*, the "dearth of studies is the legacy of a U.S. policy that considers gene-altered plants and food to be fundamentally the same as conventional ones, a policy some Americans are starting to question. . . .And it is the legacy of broken promises by the Food and Drug Administration and the Environmental Protection Agency. . ."

2. Genetic engineering may trigger allergies in people.

Genetic engineering may involve the transfer of new and unidentified proteins from one food into another, with the potential of causing allergic reactions. And allergies aren't simply a matter of slight discomfort; they can potentially result in life-threatening anaphylactic shock.

Without labeling, people with allergies won't know whether they are eating foods that contain genes from other foods to which they are allergic. In 1996, scientists were stunned to learn that soybeans engineered to include protein-rich genes from the Brazil nut contained the allergenic properties of the Brazil nut. Animal studies had not revealed the allergenic nature of the new soybean. The manufacturer halted the release of the soybean barely in time.

3. Genetic engineering may create toxins harmful to human health.

Scientists say genetic engineering may produce new toxins, with potentially devastating results for humans. In at least one case, disaster already has happened. In 1989, a genetically engineered version of tryptophan, a dietary supplement, produced toxic contaminants. Before it was recalled by the Food and Drug Administration, the mutated tryptophan wreaked havoc. Thirty-seven Americans died, 1,500 were permanently disabled, and 5,000 became ill with a blood disorder, eosinophila myalgia syndrome.

4. Genetic engineering may lead to antibiotic resistance.

Genetic engineers use antibiotic "markers" in almost every genetically modified organism to indicate that the organism has been successfully engineered. Scientists believe these antibiotic markers may contribute to the decreasing effectiveness of antibiotics against diseases.

Threats to the environment

When biotech corporations boast that genetic engineering can do wonders for the environment, we would do well to consider the source. After all, some of these companies were behind the development of such deadly pesticides as DDT. These pesticides, it was promised, would help the environment; instead, they turned into environmental disasters.

Environmentalists have many concerns about GE foods. Here are a few:

1. The plight of the Monarch butterfly

Cornell University researchers have found that GE corn may be deadly to the Monarch butterfly. In laboratory tests in the spring of 1999, the scientists found that nearly half of the Monarch caterpillars that ate milkweed leaves dusted with GE corn pollen died within four days. The surviving Monarchs that ate the genetically mutated corn pollen were much smaller and had smaller appetites than the control Monarchs, which ate normal corn pollen or no pollen at all.

Already, GE corn is being grown on 20 million acres of American farmland, right in the heart of Monarch's migratory route between Mexico and Canada. And scientists worry that there may be additional surprising scientific discoveries down the road.

2. Increased pesticide pollution

Many of the new GE crops, such as Roundup Ready soybeans, are designed to allow farmers to spray heavier doses of pesticides on their land. These pesticides inevitably will find their way into our water and food supply, endangering humans and wildlife.

New Scientist magazine reports that many farmers that have converted to GE production use as many pesticides as their conventional counterparts, while some GE farmers now use more pesticides. And one of Britain's leading safety experts, Malcolm Kane (former head of food safety at the supermarket chain Sainsbury's), has revealed that the limits on pesticide residues in soy had been increased 200-fold to help the GE industry.

3. Genetic contamination of the environment

When Scottish Parliament member Robin Harper learned that Scottish scientists were experimenting with genetically modified salmon that grow at four times the normal rate, he was horrified, and called for a ban on all genetic engineering experiments.

"We should be extremely concerned about genetically modified fish because of the danger that they could escape into the wild," he said. "It's a similar, if not even more dangerous threat, to that we are facing with GM plants. If a GM fish escaped or was released accidentally into the wild it could never be recaptured. This fish could breed with wild populations and devastate the existing natural balance with its modified behavior.

Like Harper, many scientists are concerned about the widespread release of genetically modified organisms (GMOs) into the environment. In the United States, millions of acres of land have been planted with GE crops. Scientists fear that GMOs will be spread, by bird, insect or wind, to non-GE crops—and to the wilderness. And unlike other kinds of waste, genetic contamination cannot be cleaned up, or contained.

4. Herbicide resistance and fears of the rise of superweeds

Some scientists fear that the extensive planting of genetically engineered crops will lead to a new class of "superweeds" that are resistant to pesticides. The largest class of genetic engineered foods is pesticide-resistant crops, such as Roundup Ready soybeans. The problem is that newly created transgenes may be spread unintentionally from target crops to related weed species. The weeds then also pick up resistance to the pesticide.

5. Risks to biodiversity

The terminator tree farms described on page 6 highlight a growing concern among scientists: the threat genetically engineered crops pose to biodiversity. Scientists estimate that the world has lost 95 percent of the genetic diversity present in agriculture 100 years ago. GE crops are developed from the same monoculture varieties that giant agribusinesses have planted in the latter half of this century, and will only exacerbate the problem.

Moreover, pesticide-resistant crops will allow the application of increasing amounts of powerful pesticides. These pesticides often kill more than the targeted weeds; they frequently kill beneficial plants outside their intended range.

6. The problem of unintended consequences

Biotech firms assure us there's nothing to worry about. Genetically engineered foods, they say, will save the environment.

But it's a story we've heard before. In the mid-1900s, giant agribusinesses took the technology that developed biological and chemical weapons for two world wars and used it to develop pesticides and herbicides. They promised a wondrous new agricultural era of bigger yields and bug-free produce. It was only decades afterwards that scientists began to realize the scope of the environmental devastation wrought by the explosive growth of the pesticide industry.

The discovery that genetically engineered corn might be deadly to Monarch butterflies came as a shock to biotech advocates. If biotech companies continue with their massive experiment, what will our scientists tell us 50 years from now?

Organic foods at risk

One of the best (and only) ways to avoid genetically engineered foods in the United States is to eat organically grown food. Organic foods are regarded by many people as more nutritious and delicious than their non-organic counterparts. Unfortunately, GE foods are creating a number of problems for organic growers.

1. Genetically engineered crops may contaminate organic fields.

Organic growers have warned for a long time that is impossible to avoid seed and pollen pollution from genetically engineered crops. After all, wind carries seeds, and bees can carry contaminated pollen to fields three miles away.

In 1999, their fears were confirmed when Terra Prima, a Wisconsin organic chips exporter, was forced to destroy 87,000 bags of chips at a cost of $147,000. A European importer discovered that they were contaminated with genetically engineered corn.

"Organic agriculture and genetically modified farming have both been growing rapidly. The collision of the two is inevitable," says Katherine DiMatteo, head of the Organic Trade Association. "We will probably as an industry begin lobbying for more regulations because this problem is developing so rapidly."

2. GE crops threaten one of organic farmers' most important tools.

Organic farmers do everything they can to eliminate pesticide use. But sometimes they use Bt (Bacillus thuringiensis) as a last resort. Bt is a naturally occurring pesticide that is considered to be less harmful than most manmade pesticides.

Biotech companies have created genetically engineered potatoes and corn that produce Bt in every cell. Now millions of acres of these crops have been planted. Scientists fear that the pesticide will lose its effectiveness through overuse, and that pests will develop resistance to it. Organic growers will have lost one of their weapons of last resort.

3. Biotech companies have shown a desire to tap into organic markets.

Organic activists remember 1998 as the year commercial interests attempted to squeeze genetically engineered foods, irradiated foods and foods grown in toxic sewage sludge into the definition of "organic." The U.S. government proposed including these kinds of foods in a new "organic standard." It was only after hundreds of thousands of Americans—one of the biggest activist efforts in years—wrote letters and petitioned the government to keep organic foods pure that officials dropped GE, irradiated and sewage sludge foods from the definition.

Many activists believe that biotech companies wanted to be included in the organic definition so they could tap into a burgeoning market that is growing at 20 percent per year.

Isn't the government supposed to protect us?

You might think that the U.S. government would do everything in its power to ensure that genetically engineered foods are safe for consumers and the environment. After all, the government is supposed to protect its citizens. However, this thought would be naive.

In the case of biotechnology, the U.S. government is acting more on behalf of wealthy and powerful special interests than for the common citizen. Here's what is going on:

The U.S. Department of Agriculture has acted like a cheerleader supporting Frankenfoods.

The USDA, under the Federal Plant Pest Act (FPPA), has the responsibility of overseeing genetically engineered crops. Companies that want to commercialize crops must petition the USDA. But many critics believe that USDA's oversight is insufficient, and that risky GE crops are going to market without sufficiently rigorous safety testing.

Agriculture Secretary Dan Glickman, in fact, has come across as one of biotech's biggest fans. . . instead of an unbiased protector of the people. For example, he has called the European Union's resistance to GE foods "culturally biased" and "scientifically unfounded," and has urged the EU to recognize the legitimacy of the "Second Green Revolution." In a June 1997 speech, he argued that the world must accept the American view that genetic engineering is safe and a critical piece in preventing world hunger.

The FDA ignored warnings about the safety of GE foods from their own scientists.

The FDA has consistently maintained that GE foods are safe, even though many scientists disagree. In June 1999, the Alliance for Bio-Integrity—one of the parties in a lawsuit against the FDA to force the agency to label GE foods— received internal FDA documents that show that some of the FDA's **own** scientists have doubts about the safety of GE foods.

According to the Alliance, "so strong was the FDA's motivation to promote the biotech industry that it not only disregarded the warnings of its own scientists about the unique risks of gene-spliced foods, it dismissed them and took a public position that was the opposite."

Up in arms: The world reacts to "Frankenfoods"

Genetic engineering has run into a major hurdle around much of the globe: strong public opposition. European citizens routinely tear up GE crops. European supermarkets remove genetically engineered foods from their shelves. Third World farmers rally against GE technology. And people in many countries around the world have successfully lobbied for labeling of GE foods. The response in the United States has been more subdued. But efforts here are picking up steam.

A quick look at some of the global opposition to genetically engineered foods:

GM crops uprooted in Europe, and now America

In what some claim are acts of nonviolent civil disobedience and others call vandalism, protesters have torn up dozens of GE plots in England and other European nations to protest the rise of biotechnology. In Britain, activists ripped several acres of genetically modified rape seed from a farm, and set up a flagpole and 20-foot scaffolding tripods to continue their protest. In France, protesters destroyed a small parcel of genetically mutated rape seed in southwest France.

United States farmlands have also begun to feel the wrath of GE opponents. In Vermont, for example, people cut down a 50-square-foot section of corn plants, and left three large, brightly colored cutouts of Monarch butterflies in their place (Cornell University laboratory tests last year showed that GE corn may be deadly to the Monarchs). Several similar incidents have occurred around the country.

Third World countries speak out against GE foods

In the spring of 1999, 500 farmers from India and other south Asian nations caravanned to Europe on a one-month tour to protest globalization, corporate rule and GE foods. At a protest in Britain, they said farmers in the developing world neither want nor need GE technology. Third World farmers have been particularly outspoken against "terminator" seed technology, which would force millions of farmers to buy seed from biotechnology corporations year after year.

European Supermarkets race to eliminate GE foods from their shelves

Throughout the first half of 1999, supermarket chains throughout Europe raced to remove all GE ingredients from their stores. Most major chains in Britain, and many of the biggest chains throughout the rest of Western Europe, no longer sell GE goods. In Britain, restaurants and pubs must now indicate any items on their menu that are made with GE ingredients (failure to comply can bring fines of up to $8,000).

Gerber declares it will keep its baby food GE free

The U.S. baby-food giant, Gerber, shocked food analysts and activists in August 1999 when it announced that it would no longer use genetically engineered ingredients in its baby foods. Gerber also announced that it would replace GE corn with organic corn.

The announcement was all the more surprising considering that Gerber is owned by Novartis, one of the world's largest companies involved in GE food until it announced in September 1999 that it was getting out of the business.

The move, which means Gerber is abandoning some of its long-standing corn and soy bean suppliers, will increase costs for the company—both in broken contracts and the purchase of more expensive organic ingredients. But given the emotive nature of baby food, Gerber decided the move was worth it.

Al Piergallini, president of Novartis's U.S. consumer health operation, said: "I have got to listen to my customers. So, if there's an issue, or even an inkling of an issue, I am going to make amends. We have to act preemptively."

Americans increasingly call for labeling

In the summer of 1999, Mothers for Natural Law and other groups submitted petitions to Congress with 500,000 signatures calling for labeling of genetically engineered foods. The Campaign to Label Genetically Engineered Foods has encouraged thousands more to write letters to members of Congress as well as other government officials. Surveys show that while a majority of Americans aren't aware of the issue, when they are informed, they strongly support labeling.

Why labeling?

With so many health, environmental and ethical considerations surrounding genetically engineered foods, it seems only prudent to encourage labeling as a means of helping consumers to make informed choices. The Campaign strongly supports your right to know whether the foods you are eating have been genetically mutated.

Labeling efforts already are under way in European nations, Japan, Australia and New Zealand. If these countries offer their citizens labeling protection, shouldn't the United States be doing at least as much?

There are many compelling reasons why genetically engineered foods should be labeled:

Labeling will foster consumer awareness of genetically engineered foods.

The businesses behind bioengineering have rapidly infiltrated the supermarkets with genetically engineered products; today, an estimated two-thirds of products on supermarket shelves contain GE ingredients. But so far, a majority of Americans aren't even aware that most of the foods they are consuming are genetically engineered.

The bioengineering companies have flourished under this secrecy. Only one-third of Americans are aware that their foods contain GE ingredients. The giant agribusinesses have taken over our food supply without us knowing about it. Labeling will promote a healthy public debate over the benefits and drawbacks of genetically engineered foods.

Labeling will protect people who have allergies.

Millions of Americans are allergic to certain foods. Genetic engineering may increase the risk that they will accidentally consume foods that contain allergens. If genes of a particular kind of nut are inserted in a vegetable, for example, a consumer who is allergic to that nut may be at risk. Without labeling, consumers will have no way of protecting themselves from hidden allergens.

Labeling will give people the option of whether or not to support the genetic engineering industry.

Due to all of the health, environmental and ethical questions revolving around GE foods, many people would prefer not to support the genetic engineering industry. So far, the only options they have are to buy all organic foods or grow their own food. Labeling would help people make a choice consistent with their values.

The Campaign to Label Genetically Engineered Foods

What you can do about genetically engineered foods

Sure, it's possible to feel overwhelmed by the scope of genetically engineered foods. It's a broad and complicated issue with far-reaching (and sometimes unknown) impacts, and it's going to take lots of people working together to turn the tide.

Fortunately, there's a growing movement in the United States to bring sanity to the matter of genetically engineered foods, and there are many ways you can get involved.

1. Fill out the letters on the following pages.

Sending letters to Congress and other political leaders is one of the ways you can make the biggest difference. Fill out our form letters, and encourage your friends and family to do the same.

2. Join The Campaign.

By joining The Campaign to Label Genetically Engineered Foods, you'll support one of the strongest voices for food sanity in the nation. Your membership dues also will help us reach more people with our educational information. You'll also receive The Campaign's periodic newsletter, which is full of insightful news and features about the issue.

It's easy to become a Campaign member!
Just turn to the next page, fill out the form and send it to us.

3. Pass on information to friends & family members.

We encourage you to share what you learn with co- workers, people in community groups, friends, family, and anyone else you can reach. The clear majority of Americans, when they learn about the issue, want genetically engineered foods to be labeled. The problem is, a lot of Americans don't know much yet about genetically engineered foods. That's where you can make a huge difference.

4. Keep yourself informed.

Stay in touch with The Campaign. Visit our **News Updates** page, at **http://www.thecampaign.org/newsupdates/index.htm**, to keep current with the latest news. You can also sign up on one of The Campaign's two e-mail lists, at **http://www.thecampaign.org/elists.htm** to receive news updates.

5. Buy organic.

If you want to avoid eating genetically mutated foods, one of the best ways is to buy organic. The organic food industry is growing at a whopping 20 percent or more per year, and many regular supermarkets as well as health food stores now carry organic produce and other products. Organic foods are free of genetically engineered organisms, and generally are grown without the use of pesticides.

genetic engineering

Genetic Engineering at a Historic Crossroads

The Sierra Club Genetic Engineering Committee Report April 2000, revised March 2001

Note: a list of definitions of important terms follows this report.

The last four years of the twentieth century witnessed the most rapid adoption of a new technology in history. Since 1996, millions of acres of farmland have been planted with genetically engineered (GE) crops—mainly corn, soybeans, and cotton. This means that genetically engineered organisms (GEOs) are being released to the environment on a massive scale, an event unprecedented in the 3.8 billion year history of life on this planet. This technological upheaval happened virtually without public debate, while our government played the role of enthusiastic promoter, rather than cautious regulator, of this radically new and environmentally hazardous technology.

Genetic engineering is a new technology that combines genes from totally unrelated species, in combinations not possible using conventional breeding methods. Genes from an animal, say, a fish, can be put into a plant, a strawberry for instance. In fact this is an actual example of an attempt to "improve" strawberry plants. The fish gene is supposed to make the strawberries more resistant to frost by causing the strawberry plant to produce a form of antifreeze which the fish normally produces to endure cold ocean conditions.

Over 60 percent of all processed foods purchased by U.S. consumers are manufactured with GE ingredients. Some corn and potatoes have even been genetically engineered to contain a gene from Bt bacteria which causes every cell of the plants to produce an insecticidal toxin. Yet there is no labeling of these or any GE foods as being genetically engineered, because the U.S. Food and Drug Administration (FDA) considers the GEOs from which these foods are made to be "substantially equivalent" to the non-genetically engineered plant from which the GEOs are derived.

The doctrine of substantial equivalence is pure pretext and rationalization with no basis in science. Yet, in a remarkable display of arrogance, the supporters of genetic engineering accuse their critics of not basing their objections on "sound science."

The FDA also uses this "substantial equivalence" rationalization as an excuse to avoid any effective testing of GE foods to determine their safety. Such testing might seriously delay or even prevent the introduction of GE crops into the marketplace.

From the time in the early 1970s when advances in molecular biology led to the development of the techniques we call genetic engineering, until the mid 1990s, the organisms produced by genetic engineering were nearly all confined to laboratories or controlled factory settings. During this time there were almost no releases of genetically engineered organisms to the environment, as genetic engineering was used in basic research and to produce medically useful substances such as insulin.

The unrestrained expansion of genetic engineering into agriculture during the past four years changed all that. By 1999 almost 80 million acres of North American farmland were planted with genetically engineered seed. This means massive releases of GEOs to the environment are now taking place. Genetic engineering now poses a very grave threat to the natural environment.

Historic turning point

We are now at a turning point in history. We can continue to allow the virtually unrestricted release of genetically engineered organisms to the environment, or we can bring this technology under strict control.

If we continue on our present path of unrestricted releases of GEOs, we will eventually live in a genetically engineered world, as the genome of each species now on earth is either deliberately altered by genetic engineering or indirectly altered by inheritance of transgenes from a genetically engineered organism. In such a world there would be nothing left of living nature, as every species would have been deprived of its genetic integrity, and every ecosystem would thereby have been irreversibly disrupted.

Of special concern to environmentalists should be the fact that trees are now being genetically engineered, and that it is proposed that entire forests be planted with these trees. One such proposal is for trees which produce no seeds, but divert the energy from seed production to more rapid growth of wood. A forest of such trees would wreak havoc on the food chain. Other GE trees that do produce seeds could cross with native varieties and damage forest ecosystems. Engineered trees which produced pollen (as might happen despite scientists' attempts to create sterile subspecies) could cross with native varieties miles away and damage forest ecosystems.

Fish, as well as other animals, are also being genetically engineered to grow more rapidly. If they are released to the environment (fish culture tanks often discharge during storm conditions), they may out-compete native species and thereby disrupt ecosystems.

There is evidence that soil organisms may be adversely affected bygenetically engineered crops. The Bt corn plant is engineered to contain a bacterial gene that causes production of an insecticide in every cell of the corn plant, including the edible corn ear and the roots. This toxin has been found to persist in the soil for months.

The promoters of genetic engineering show no sign that they are willing or able to impose limits on their applications of this new technology. It will therefore be left to the institutions of civil society—governments working with non-governmental organizations representing the concerned public (such as Sierra Club)—to set limits to how much further genetic engineering will be allowed to alter the earth's species.

Medical uses

Promoters of the use of genetic engineering outside the laboratory claim that a moratorium or other controls on the planting of genetically engineered organisms as agricultural crops would mean an end to the uses of genetic engineering in the production of medically useful products. This is untrue. As long as proper precautions are taken to assure that the genetically engineered microorganisms used in production of pharmaceuticals or in scientific experiments are not released to the environment, such uses need not be prohibited. However, all applications of genetic engineering, including medical uses, carry some risk. Medical applications of genetic engineering should be approached with caution and not rushed to market.

We believe that simpler, more traditional strategies for problem solving should always be considered when evaluating the production of transgenic organisms. This is especially relevant with respect to agricultural applications, as will be discussed in the topic below.

Feeding the world's hungry?

Medical uses of genetic engineering may be prudent, but agricultural applications of GE are not. Yet the argument is being made by the biotech industry that if the genetic engineering of farm crops is not allowed to proceed, the poor people of the world will starve.

In fact there is more than enough food produced by conventional agriculture, without genetic engineering, to feed all of the world's people. One cause of hunger is the ineffective distribution of food. Genetic engineering may actually lead to more food insecurity and hunger because in poor countries it will lead to the planting of monoculture crops, highly vulnerable to disease and pests, in the place of resilient, diverse range of crops, and it will make farmers dependent on corporations that will demand payment for basic inputs such as seed, chemicals, and fertilizers.

Terms of trade between developed and less developed nations have often resulted in the best land in the poor countries being used to grow cash crops for export rather than food for consumption at home. Issues of equity and fairness have not been addressed by trade agreements. Certainly these problems call out for redress, but their solution isn't to increase the monopoly power of "life science" companies in the richest nations.

As Indian writer and activist Vandana Shiva summarized, "Millions of farmers in third world countries want to breed and grow the crop varieties that adapt to their diverse ecosystems. Plant biodiversity is essential for a balanced diet. Yet numerous crops are pushed to extinction with the introduction of GE crops."

Terminator technology

Any claim by the corporations promoting agricultural biotechnology that they have the interests of the world's poor people at heart are refuted by the facts in the case of Terminator seed technology. This technology would protect the intellectual property interests of the seed company by making the seeds from a genetically engineered crop plant sterile, unable to germinate. Terminator would make it impossible for farmers to save seed from a crop for planting the next year, and would force them to buy seed from the supplier. In the third world, this inability to save seed could be a major, perhaps fatal, burden on poor farmers. The Sierra Club's Genetic Engineering Committee (GEC) believes that Terminator is a tool by which seed companies are trying to engineer their monopoly power into the genetic code.

Adding insult to injury in the Terminator technology story is the fact that our own tax dollars were used to develop Terminator. The U.S. Dept. of Agriculture played a major role in the development of Terminator technology. The USDA is actually part owner of the patent on this immoral technology.

The Genetic Engineering Committee

The Sierra Club's Genetic Engineering Committee (GEC) was formed in May 1999 to explore ways to mobilize the strength of Sierra Club, the largest grassroots environmental organization in the U.S., for the work of public education and regulatory reform that will be necessary to protect the natural environment and human health from the threats posed by the release of genetically engineered organisms.

An Educational Challenge

Genetic Engineering Committee members have found that the need for public education is great. Because of inadequate reporting by the U.S. media, many otherwise well educated people simply have not been told what genetic engineering is. We hear statements like, "If there is a moratorium on planting genetically engineered crops, doesn't that mean that no crops at all will be planted?" And, "Aren't all farm crops these days genetically engineered?"

The supporters of genetic engineering gladly fill this information vacuum with false statements. They claim that the selective breeding of plants and animals that has been done for centuries is genetic engineering. Supporters claim that modern genetic engineering is nothing more than an improved, more precise, high-tech form of conventional plant and animal breeding. Michael Khoo, in a letter published last year in the Toronto Globe and Mail, called this claim ". . . biotechnology's public-relations line that genetic engineering is no different from traditional breeding." His letter continued, "A potato can cross with a different strain of potato but, in 10 million years of evolution, it has never crossed with a chicken. Genetic engineering shatters these natural species boundaries, with completely unpredictable results. As a result of these risks, the British Medical Association has recently called for an open-ended moratorium on GE planting."

Gene transfers occur in conventional breeding, but these transfers can only take place between individuals of the same species, or, in the case of hybridization, between individuals of closely related species. This is because conventional breeding relies on the normal reproductive processes of the plants or animals. Plants can be conventionally bred only with plants of the same species or, to make a hybrid, with closely related species. And animals can only be bred with other animals of the same or, in some instances, closely related species.

Genetic engineering is not bound by these limits in the possible exchanges of genes that can be made to occur using its techniques, which include the use of viruses as "vectors" to move foreign genes into host organisms. By means of genetic engineering, genes can be transferred from a plant to an animal; from an animal to a plant; from a bacteria to a plant, and between numerous other combinations of donor and recipient organisms. There have even been attempts to put human genes in plants and animals that are used as human food.

Why is this important?

The changes caused by genetic engineering can be inherited by subsequent generations of the affected organism, and, once released to the environment, these organisms cannot be recalled—they will continue to pass on their spliced-in genes, or transgenes, to future generations. Many of the gene changes may turn out to have unexpected secondary effects. Serious errors in judgement might prove unrecallable as trillions of copies are broadcast via pollen and seed. Wild relatives of crops will also be affected, with possibly profound effects on the environment. For instance, genetically engineered cereals may cross with various grasses. Once this process begins, it will be for all practical purposes uncontrollable and unpredictable.

Biodiversity and endangered species

As environmentalists, one of our most basic concerns is the preservation of species. We live in a time when the rate of species extinction has increased drastically, primarily as a result of human activities. Now a new form of human activity, genetic engineering, may pose the ultimate threat to the survival of all species.

Many of those who are promoting genetic engineering give every indication that they regard life as a form of information technology: that genes are mere bundles of information to be transferred from one species to another on the basis of expediency and potential corporate cash-flow; that the natural barriers to genetic transfer that protect the integrity of species are mere inconveniences to be overcome; and that the very concept of species is an anachronism which it is now time to discard.

Because these principles are being put into application—genetically engineered organisms are now being made and released to the environment—we have to conclude that genetic engineering threatens the continued existence of all species as life-forms that are distinct from one another.

Genetic engineering should be considered an environmentally dangerous technology that is breaking down the barriers that have protected the integrity of species for millions of years. There are probably good reasons why it is impossible for a conventional plant breeder to combine plant genes with animal genes. Those reasons have to do with the very survival of life on earth, and we ignore them at our peril.

Another threat to biodiversity from genetic engineering is from toxins produced by GE crops. The finding last May that Bt corn pollen might be a threat to monarch butterflies provides an example. Genetically engineered Bt crops have the gene spliced-in from the Bt bacteria that codes for the production of the toxin that kills insect larvae. A Cornell study showed that this toxin kills the larvae of certain species of moths and butterflies. Other studies have indicated reduced viability of other nontarget beneficial insects, such as ladybugs and lacewings. Bt toxin also persists in the roots of the crops and in plant residues for a considerable time after the crop is harvested, which may have major adverse consequences for the millions of soil organisms that help maintain soil fertility.

Yet another threat to biodiversity is the out-crossing of herbicide resistance traits to native plants. There is already evidence of "superweeds" created by the spread of pollen carrying the herbicide resistance trait.

A threat to organic farming

The standards established by organic farmers categorically exclude genetically engineered crops from the organic food system. A problem arises from pollen drift from fields of GE crops planted too close to organic crops. The organic plants may become crossed with the GE plants and thereby contaminated with the spliced-in gene (transgene) from the GE crop. Then the crop grown next season from seed saved from what was an organic field will no longer be acceptable as organic—it will contain the transgene and will have to be considered genetically engineered. And in the case of crops in which the harvested portion of the plant is the seed, the presence of a transgene will immediately, in the first generation, make the crop not acceptable as organic. This problem of outflow of transgenes to organic crops is considered by organic growers to be very serious.

Another negative impact on organic farming is the expected resistance that insect pests will develop to Bt toxin. Organic farmers have been using Bt bacteria applied to crops in a spray as an organic method of controlling damaging insects. But genetically engineered Bt crops have the gene that codes for Bt toxin production spliced-in. By applying Bt bacterial sprays only occasionally, and because of the naturally limited quantity of the toxin present in the bacteria, organic farmers have avoided pest resistance problems. Now, with massive quantities of Bt toxin present in fields throughout the growing season, most of the insects susceptible to the toxin will be killed off, leaving a proportionately greater number of resistant insects alive. These Bt-resistant survivors will pass resistance traits into future generations. It is expected that resistance problems caused by genetically engineered Bt crops will render Bt sprays useless to organic farmers within a few years.

Health issues

While Sierra Club is an environmental organization, we are concerned also with potential human health impacts of this new technology. Among the issues are the possible spread of allergens, the invitation which herbicide tolerant crops give to over-use of herbicides, possible adverse effects of new toxins (such as the Bt endotoxin) on some people, and the emergence of antibiotic resistance which may be fostered by the use of antibiotic resistance genes in almost all transgenic crops. New genes also alter the expression of native genes and so may change the nutritional benefits of foods and may also result in the overproduction of previously low-level natural toxins which exist in most foods. Health risks add to the environmental reasons for exercising caution.

The Precautionary Principle

The Genetic Engineering Committee strongly supports application of the precautionary principle to biotechnology issues and recognizes the limits inherent in present systems of risk assessment. Here is cogent statement of the precautionary principle from the Wingspread Consensus Statement on the Precautionary Principle, Jan, 1998: "When an activity raises threats of harm to the environment or human health, precautionary measures should be taken even if some cause and effect relationships are not fully established scientifically."

The participants at the conference said the following about risk assessment: "We believe existing environmental regulations and other decisions, particularly those based on risk assessment, have failed to protect adequately human health and the environment, the larger system of which humans are but a part."

Carolyn Raffensperger offered further commentary on risk assessment: "Participants [at the Wingspread conference] noted that current policies such as risk assessment and cost-benefit analysis give the benefit of the doubt to new products and technologies, which may later prove harmful. And when damage occurs, victims and their advocates have the difficult task of proving that a product or activity was responsible." (email by Ms. Raffensperger, 1/28/98)

The precautionary principle is of the greatest importance when the damage from a new technology would be irreversible. This is the case with genetic engineering. Once they are released into the environment, genetically engineered organisms cannot be recalled. The Genetic Engineering Committee believes that genetically engineered farm crops are wrongly given the benefit of the doubt in the regulatory process, and that, under the precautionary principle, they should not be released into the environment or allowed to be part of the food supply.

The regulatory process

The federal government decided early in the development of genetically engineered crops that this was a technology where U.S. producers had an advantage which could be used to help them compete successfully in world markets. It was decided during the first Bush administration that the regulatory process for approval of GE crops would be streamlined. The Clinton administration continued this policy, with both President Clinton and Vice President Gore being strong supporters of agricultural biotechnology.

The regulatory inadequacies in the case of Bt potatoes are illustrative. The U.S. Food and Drug Administration (FDA) does not test the toxin in Bt potatoes for safety as a food additive because the toxin is a pesticide and therefore the U.S. Environmental Protection Agency (EPA) has the responsibility to assure its safety. But the EPA tests only the Bt toxin, not the potatoes containing the toxin. So no one tests Bt potatoes for their safety as food, yet they become part of our food supply. The FDA does not require labeling of Bt foods, because the agency is prohibited from requiring any information about pesticides on food labels, and because they consider GE foods to be substantially equivalent to conventional foods. Meanwhile, the U.S. Dept. of Agriculture pursues a role primarily of promotion of genetic engineering in agriculture, spending only a tiny fraction of its budget on safety testing of biotech foods.

As for testing GE crops for environmental hazards, there has been no environmental impact statement (EIS) done for a release of any genetically engineered crop. This is in violation of the National Environmental Policy Act (NEPA), which the EPA has the responsibility to administer.

Proposed Legislation

Laws are needed to require safety testing and labeling of GE crops. Also needed are mandatory environmental impact statements for every ecosystem into which any new GEO is to be introduced, and when applicable, involvement with the U.S. Fish and Wildlife Service. Liability issues also need to be addressed: clarification is needed as to who is responsible for the downstream effects of a company's product, including damage to organic producers and damage to the environment. Funding for agricultural research and development should be directed towards sustainable methods, rather than methods that perpetuate dependence on the chemical treadmill and agricultural biotechnology.

Moral and Religious Issues

In the book *My First Summer in the Sierra*, by the Sierra Club's founder, it becomes more clear with each page that John Muir regarded the study of nature as an act of worship. Although Muir provides glorious images of inanimate nature in the Sierra—the mountains and rock formations; the clouds, storms, and waterfalls—most of the book is devoted to careful descriptions of plants and wildlife, with frequent mention of how all these living things are loved by their Creator. Muir tells us that to be in a place like the Sierra Mountains is to be closer to God than is possible in any human-built church.

Not all Club members will hold the same religious convictions that John Muir held. But most of us probably share his belief that ethical principles are a part of our relationship with nature; that there is a moral dimension to our task of protecting nature. Those ethical principles lead us to respect and protect the natural world.

To Muir, the more one knows about nature, the more one is inclined to protect nature. Using our knowledge of living organisms in order to better protect wild nature is the opposite of using that knowledge to bring living things into the realm of human technology and human control.

There is not a shred of evidence that John Muir would have regarded the release of genetically engineered organisms to the environment with anything but shock and outrage. We can be certain that Muir would be fighting those powerful corporate forces that are trying to control and commodify the very basis of life. We can be certain that Muir would commit to this fight the same energy and spirit that he gave to his last battle, the struggle to save the Hetch-Hetchy Valley. We believe that Sierra Club today should commit major resources to save what remains of living nature from this new technology of genetic engineering.

Respectfully submitted,

The Sierra Club Genetic Engineering Committee

What you can do—
Write to your members of Congress urging them to support the bills discussed above.

Start biotechnology committees within your own region.

Make sure the public understands what genetic engineering is. If a statement appears in the media repeating the myth that genetic engineering is nothing more than what conventional plant and animal breeders have been doing for centuries, write a letter to the editor stating that genetic engineering is a new and dangerous technology that combines genes of unrelated species. See letter writing tips.

Definitions of key terms:
Biotechnology - A term now widely used to mean genetic engineering. In a larger sense, biotechnology is any use of biological processes to produce a desired result. Thus, the use of yeast to bake bread is a form of biotechnology which is not genetic engineering and which has been in use for centuries.

Genome - The complete set of genes of an individual organism, or the complete set of genes of all the individuals of a species.

Genetic engineering (GE) - The transfer of genes from one organism to another organism in ways that are not possible using conventional breeding methods. Genetic engineering bypasses the reproductive barriers that prevent genetic transfers between unrelated species, thus allowing transfer of genes from an organism of one species to another, completely unrelated species. Genetic engineering also includes methods of gene deletion and gene manipulation that are not possible using conventional breeding methods.

Genetically engineered organism (GEO) - Any living thing that has had its genetic structure altered by genetic engineering. A genetically engineered organism is also called a genetically modified organism (GMO), a genetically altered organism, or in certain cases, a transgenic organism.

Recombinant DNA technology - The technique, also called gene splicing, that made possible the first application of genetic engineering, in 1973. A section of DNA molecule which constitutes a gene, the basic unit that determines an inherited trait, is cut from the molecule and spliced into another DNA molecule in another organism. The two organisms need not be of the same species or even closely related. Thus, using recombinant DNA techniques, genes from bacteria have been spliced into corn plants and DNA from a fish has been spliced into strawberry plants. It is also possible to splice plant DNA into an animal.

Transgene - A gene from one organism transferred into another (usually unrelated) organism by means of genetic engineering.

Transgenic organism - An organism containing a transgene.

Vector - In the context of genetic engineering, a virus or plasmid used to transfer genetic material into a cell.

From The Sierra Club, "Genetic Engineering at a Historic Crossroads," Biotechnology Task Force Report, April 2000, revised March 2001. With permission.

---

http://www.sierraclub.org/biotech/report.asp

EATING IN THE DARK

Text of Our *NYT* Op Ad

The TomPaine.com Staff .

FDA Will Not Require Labeling of Genetically Engineered Foods

Americans have a right to know what's in our food.

So how come the Food and Drug Administration wants us eating in the dark?

The FDA has proposed new rules that would not require genetically engineered food to be labeled as such. The rules would also continue to allow these foods to be sold without any required safety testing.

Very little independent research has been published on the safety of genetically engineered (GE) foods. The FDA's own scientists have warned that there's not enough evidence to declare them safe. Yet, in what amounts to an uncontrolled human experiment, the FDA has already allowed GE foods to become part of our diet.

We don't know what these foods might do to people with allergies or weak immune systems, or if they have any long-term effect on children. Biotechnology companies might know, but in the name of protecting trade secrets they have kept most of their test results private and away from peer review.

We do know this: Credible polling shows consumers overwhelmingly support GE food labeling. Yet the FDA has ignored the public's desire, proposing rules that give the biotech industry just what it wants. And no wonder. Generous contributions to both political parties give the industry special access to FDA's overseers in Congress and the White House.

The new FDA rules are not yet final. Consumers have one more week -- until May 3 -- to let the agency know what they think. They can do so through the website www.TrueFoodNow.org.

The 15-nation European Union, Japan, Australia, New Zealand, South Korea and Russia all mandate the labeling of genetically engineered food.

But if the FDA's new rules go through as drafted, Americans will be left eating in the dark.

This Week at TomPaine.com -- Eating in the Dark

Featuring a detailed critique of the FDA's proposed rules ... "The A-B-C's of GE Food" by Rachel Massey ... and "Common Sense on Biotech" by Michael F. Jacobson.

GE brings new things to life

LIKE TERMINATOR TECHNOLOGY AND FRANKENFOODS

The GENETICALLY ENGINEERED FOOD fight

by Phillip Frazer

Odds are that you are eating food made from genetically altered (GA) plants, every day, in every meal. The brave new world of GA plants has arrived, and its backers plan a vastly expanded menu of items: from tomatoes that kill bugs to lemon-scented lawns.

Leading the biotech charge are four big "life science" corporations: Monsanto, Novartis, Hoescht Chemical and Du Pont (they used to be called chemical companies).

Americans are buying GA food in huge quantities, almost entirely unaware that they are doing so.

The Europeans, on the other hand, are putting up stiff resistance (see *NoE*, August 1998). A British poll found 96% of respondents wanting labels on GA foodstuffs. Critics there call them "Frankenfoods." The European Union must approve each GA plant before it can be grown or marketed and so must each member government. Austria and Luxembourg have outlawed GA food outright.

DO NOT ALTER

Slogan of opponents of genetically altered foods, also known as genetically engineered (GE) or genetically modified (GM). GE, GA or GM plants are GMOs, or genetically modified organisms, or GEOs or GAOs, or transgenic.

Almost all of the 30 million acres planted worldwide with modified seeds in 1997 were in the US, including 30% of our soy, 25% of corn, 40% of cotton, and 50% of canola.

What are they?

A little background: every cell in a plant has two

What are they?

A little background: every cell in a plant has two

sets of chromosomes, one from each parent, that control the plant's life. These chromosomes are made up of genes, each of which is a piece of DNA that contains a code to make a specific protein. Proteins make up much of the plant's cells and also "tell" the cells what to do. "Genetic engineering," says Martha L. Crouch, associate professor of Biology at Indiana University, "can be defined as the process of manipulating the pattern of proteins in an organism by altering genes. Either new genes are added, or existing genes are changed . . . [and] because the genetic code is similar in all species, genes taken from a mouse can function in a corn plant."

Most of the alien genes that have been inserted in our crops are there to kill bugs or fungi or to allow the plant to withstand a barrage of weed-killer.

Why is this different than using seeds "created" by cross-breeding?

GA advocates say farmers and scientists have long toiled toward these goals by selectively breeding new varieties of plants and animals. Genetic alteration, they say, does the same thing. But others disagree.

Many hybrid strains are sterile, meaning they will not spread to overtake their naturally occurring relatives. GA plants reproduce as well as or better than

Many hybrid strains are sterile, meaning they will not spread to overtake their naturally occurring relatives. GA plants reproduce as well as or better than natural ones, unless they are programmed not to (see "Terminator Technology" above).

Furthermore, in hybrids the desirable traits are in their "right" location on the chromosome. In genetic engineering the desired data is shot into the chromosome at random, with unpredictable results.

"This is not an extension of classical breeding," says America's foremost critic of biotechnology Jeremy Rifkin. "[W]ith genetic engineering technology you can cross all the biological boundaries: you can make mice with human growth genes and you can have firefly genes lighting up tobacco plants."

Rifkin and many other critics worry about what these new plants will do to birds, insects, microorganisms and animals who eat them. And, says Rifkin, "it is the scale that is important." Thousands of new organisms are, right now, being introduced to our water, air and land, covering millions of acres.

Not to worry, say advocates of genetic engineering, everything will be tested thoroughly; we will find out what our new plants or bacteria do before they are let out of the lab. But that is not always the case. Not long ago scientists said there was no way that inserted genes could jump from the engineered plant to another species. But they do (see "Superweeds and superbugs," p. 4).

For the farming industry, what makes GA superior to hybridization is that a precise trait can be introduced without worrying about other genes coming from the cross-bred plant. Using recombinant DNA techniques they can introduce genes from totally unrelated organisms: the gene for resisting cold, for instance, has been transferred from North Atlantic flounder fish to potato and tomato plants, and powerful poisons that bacteria use to kill insects have been added to seeds. The bug-killing bacteria, known as Bt, is a favorite of gardeners and farmers who are wary of commercial pesticides. They worry that having that gene in seeds will hasten the evolution of Bt resistant bugs, ruining Bt as an organic alternative to chemical insecticides.

Too much to regulate

The federal government is spending a ludicrously small $1 million a year on assessment of the risks involved in GA research; essentially, the biotechnology industry is regulating itself.

"The small-scale field tests that are done on genetically altered plants have not been designed to investigate the ecological risks associated with widespread commercialization," according to Allison Snow and Pedro Moran Palma writing in BioScience (47:2). (These field tests are posted on the Internet at www.nbiap.vt.edu.)

While the corporate scientists wax reassuring, the FDA, EPA and the other underfunded government agencies only look at GA plants if they do something truly new, such as soybeans that have a much higher oil content than usual. As for long-term human studies, they are coming from the field. That means real people like you and me, eating GA food whether we know it or not, are the real-life guinea-pigs.

All of this explains why the insurance industry declines to provide long-term insurance for any of these biotech endeavors. What if a weed is cross-pollinated by accident (for example, by the wind) and acquires resistance to whole categories of weed-killer. That weed could ruin thousands, perhaps millions, of acres of crops—who pays? Not us, say the insurance companies. They will only insure short-term crop damage or negligence.

As Jeremy Rifkin says, "There is no predictive ecology. . . . There is absolutely no ecological risk assessment science by which to do it."

Rifkin says the life sciences companies are spreading genetic pollution which will come to be a greater curse than chemical and nuclear pollution have turned out to be. Ironically, this is where he takes heart. He believes the liability issue will turn agricultural biotechnology into "one of the great disasters of corporate capitalist history."

"Remember that the baby-boom generation rejected nuclear power, the crown jewel of 20th century physics," he says. "When I was growing up in the 1950s that would have been unthinkable."

Rifkin also believes that "the middle class sets the trends in Europe, Japan and North America . . . and [they are] moving towards organic foods."

The life sciences industry has chosen the hard path, creating plants that defend against the rest of nature. Rifkin argues for the soft path in which our ever-expanding understanding of the genetics and growth habits of naturally occurring plants allows us to mix varieties from farm fields with traditional flora in local ecosystems. This, say proponents of organic, local and sustainable agriculture is how we will feed the world—not with a mess of Frankenfoods.

☞ go to www.newscientist.com/nsplus/insight/gmworld to read an interview with Rifkin and more
Rural Advancement Foundation International at www.rafi.org
Physicians and Scientists Against GE Food at www.psagef.org

Section VI

Business Documents

As discussed in *Environmental Politics*, much of the advocacy for business interests is conducted by foundations and think tanks who support free-market principles and limited government. Business interests, again as suggested, have had to tread carefully as environmentalism has achieved increasing popularity.

The most interesting and compelling arguments for limited regulation with respect to bioengineered food, as with other issues, have been made by these foundations and think tanks. I have therefore included two of the most provocative and challenging. One is an extended interview conducted by Ronald Bailey with Nobel Peace Prize laureate, Norman Borlaug in *Reason*. The other is a rebuttal to those who would require labeling of such foods in Consumer Alert. *The New York Times* op-ed by Daniel J. Popeo, from the Washington Legal Foundation, is an effort to reacher a broader audience than the others. Another such argument, in Cato's journal, *Regulation,* is reproduced in the Media section of this Casebook.

Biotech companies do, of course, advance their interests on their Web sites, but, again, they frequently refer and link to others to make their case. But Monsanto has established — and prominently displays — its new pledge, which I have provided here.

Discussion

The work of corporate funded think tanks on behalf of business and industry is increasingly influential. In having them make the case for limited regulation, business seems less self-serving. Moreover, the interests of business are philosophically consistent with the market driven environmentalism that they stand for. Of course, it was a windfall to have someone with the scientific credentials of a Norman Borlaug supporting the genetic modification of food. Supporters of bioengineering generally ground their arguments on science, which has yet to find any evidence of harm, hence Daniel Popeo's direct attack on biotech opponents as "luddites." Anti-bioengineering people generally base their arguments on the "uncertainty" of risk.

Similarly, the argument against labeling in Consumer Alert suggests that "tastes and values" are being substituted for science. How persuasive is this argument? In fact, how persuasive is the argument that biotech supporters are being asked to make a difficult case, i.e., that there are no risks nor will there be, even in the absence of any adverse consequences in the decade that they have been widely consumed.

Monsanto's pledge, on the other hand, should be viewed in the context of the codes of conduct that businesses began adopting in the 1990s to assuage public concerns that they were indifferent to the environment (see Chapter 8, *Environmental Politics: Interest Groups, the Media, and the Making of Policy*). They are apparently still suffering from the overconfident public relations effort they waged in the early 1990s, which failed to anticipate the public reaction they are now getting.

Food Labeling — The Problems of Mandated Information for Biotechnology

by Frances B. Smith, executive director, Consumer Alert

(September 1999)

The argument many activists use for the expansion of mandatory food labeling information is that consumers want and need more information in our complex modern age. New food products and new processes abound — shouldn't companies be required to provide consumers with information to make decisions, especially about the new? This argument is currently being used by activists to press for mandatory labeling of food produced through biotechnology.

At first glance, that contention may sound plausible. Consumers do need and want product information. But what kind of information? How much? For whom? For what purpose? Based on what criteria? Selected and provided by whom? These are all valid questions that need to be answered before a mad gallop to mandate new label information tramples common sense. Labeling is supposed to provide information. Now, with a greater push for government-mandated labels, it seems to be a way of endorsing certain values, of favoring some cause.

In the food area, approaches to labeling can play a large role in consumers' perceptions and in their acceptance of those foods, particularly when the process used in the food production is new or the food itself is novel. Mandating that certain labeling information is needed because consumers want to know more about the food or its production presents the risk that the information will relate to some people's tastes and values rather than science.

What's on a Label

Food producers provide information on labels for a variety of purposes:

- to give specific information about the product itself, particularly if it is a new product,
- to differentiate the product from its competitors,
- to inform consumers about how to use or prepare the product,
- to warn consumers about potential problems or improper use,
- to provide manufacturer contact information, and other purposes,
- to provide a guarantee through the "brand name."

All of this information must be truthful, and if it deals with scientific claims, the underlying scientific studies have to be evaluated. So if a label claims that the food provides health benefits, studies must demonstrate that those claims are scientifically valid.

"Branding" for Quality

Consumers evaluate many factors before deciding whether to buy a product — past experience, recommendations from friends, reputation of both the producer and the seller. And, of course, information on labels can also be relevant.

To many consumers, one of the most important pieces of label information is the "brand name," that is, the name of the food producer that stands behind the product. The "branding" itself provides information that a certain quality level exists. And it is not a new concept. Craftsmen throughout history have carved their market on furniture, musical instruments, clocks, and other products. Artists signed their names to their works to identify them as authentic creations of the "master" or his school.

Consumers, through experience, aided by advertising, ratings, or word-of-mouth recommendations, learn which brands consistently provide them with the quality they desire at the price they want to pay. Since that "branding" also can extend over numerous product lines, a poor quality level in one of a producer's products can spill over to another and hurt the total brand name, and hence sales. Food companies know that and thus make sure that their foods meet their standard across the board.

Specialized Needs and Specialized Food Labels

Besides "branding" their products, food producers, on a voluntary basis, provide information important to consumers, such as how to prepare the food properly. The 800-numbers for questions about the products are popular with consumers. In many cases, particularly with new products, companies' self-interest in selling their products drives them to tell consumers on labels and in advertising about the product's attributes. Some label information — for example, the list of ingredients — is required by federal agencies.

People have very different views about what information they want on food labels. Consumers, depending on prior knowledge and their own needs and values, may find some of the information more important to them than it would be for others. They search out information that they need or value, whether it relates to ingredients, safety, diet or nutrition, or warnings. Both the product and the process used to produce the food may be important to some people — those who follow religious dietary codes or have other value-based preferences. Groups desiring specific label information relating to their tastes or values have worked with manufacturers to create specific labels and to certify certain products, for instance, kosher food.

In the private market, those requirements or preferences result in specialized companies producing food that satisfies those specialized values, and then labeling and marketing the food so that the process is clear. Thus, a company that wants to provide consumers with facts about production methods may do so, for example, in the case of organic agricultural methods or food that complies with religious codes. Producers identify and label the products that meet certain criteria, and consumers who want to purchase kosher food or organically grown food can readily find those products. In the private sector, those criteria don't have to be the same. For instance, one kosher certification label may have much stricter criteria than another. Those specialized products carry specialized information for people with special wants or needs.

New Processes, New "Warnings"

Producers may also want to provide label information to address consumer concerns about food safety. For example, a company that produces frozen ground beef patties may want to let consumers know that their patties were "cold pasteurized" with irradiation to reduce significantly the incidence of E.coli and many other food-borne pathogens. Since scientific studies support those claims, companies may want to highlight the process in their product marketing. Some consumers concerned about food-borne diseases may be attracted to "cold pasteurized" foods, while others suspicious of new technology may avoid them.

Labeling of Irradiated Food. In contrast to the voluntary provision of information, almost invariably government-mandated wording on food labels, even when designed for informational purposes and not as warnings, shrieks "Stay away from this product!" Current federal government labeling requirements for irradiated foods mandate wording stating that the food has been irradiated, and require a symbol — or radura — to depict that fact. The irradiation label in this form only serves as a warning. It is likely that some consumers would be misled into thinking that food that has been treated with radiation is less safe or that it is radioactive. The wording may in fact cause some people to shy away from such foods in that belief and risk missing out on a critical way to protect themselves and their families from food-borne diseases.

The private market can likely find many innovative ways to promote irradiation as a positive good that enhances their food products. A government mandated "one size fits all" statement and logo gives less incentive to companies to more creatively promote the value of irradiation, especially if the "one size…" is perceived as somewhat alarming to consumers.

The Olestra Example: The case of the labeling of products containing olestra — a fat substitute approved for use in salty snacks — provides a real-life example of how mandated information translates into a warning. The mandated label was not supposed to be a warning label but an informational one — to inform potential customers that some consumers may be more sensitive to non-absorption of olestra by the body and may experience unpleasant gastrointestinal effects. The alleged intent was to provide some consumers with information that would help them make a more informed decision about purchasing and using the product; it can be characterized as "Take care — you may be especially sensitive to the fact that olestra is not digested by the body."

However, the "informational" label prescribed by the government goes far beyond that. The notice on snack foods with olestra tells people in graphic detail what some of the negative effects might be. Although technically an "information" notice and not a "warning" label, the label requires language that raises the specter of disgusting, distasteful, and disagreeable consequences of eating olestra snacks. The nature of the language thus raises an alarm, rather than just providing information.

The olestra label seems to apply a new principle to this food. Many traditional foods have various impacts on some people or on some people some of the time. Drinking milk or eating dried fruit or fiber-enriched cereal can cause some people some distress sometimes. But it is not the usual practice to require those foods to have special label information. Yet special requirements were mandated for an innovative food, and those requirements were negative.

"Broadcase" vs. "Narrow" Approach

Government mandates for labels also often fail to consider the "broadcast" vs. "narrow" information issue. An information label is a "broadcast" approach — it gives everyone the same information even though that information may be relevant or of interest to only certain people. Is that the best approach or would a more "narrow" approach be more effective?

For example, lactose intolerance is a condition affecting many people, usually not that seriously. Yet foods containing lactose do not carry a specific information label going to the general public that notes that fact or points out the risk/discomfort of lactose for those who are intolerant of that ingredient. Rather, the minority of people who are sensitive to lactose seek out lactose-free, acidophilus milk and other products.

"Crowding Out"

An overload of mandatory label information can also "crowd out" essential information on food products. Too much "clutter" or "noise" on labels makes it more difficult for consumers to locate important facts about products. That "crowding out" problem in which mandates for prominent label information may obscure critical information can actually harm consumers in interfering with their ability to find that information. In the case of a person with an obesity problem, the caloric value may be more important, or, for those on sodium-restricted diets, that information would be more valuable.

In the case of many food products, critical information about food storage, how to cook the food properly, whether the food needs refrigeration after opening is less visible to consumers and thus may be overlooked. Yet lack of knowledge about proper cooking and storing can greatly increase the risks of food-borne diseases for many consumers. Too much information often means too little — in the sense that little information is read or important information is overlooked or cannot readily be found.

Research findings of a Consumer Labeling Initiative by the Environmental Protection Agency (EPA) found that very problem: Many consumers were not reading the mandated product labels for household insecticides and pesticides under the Federal Insecticide, Fungicide, and Rodenticide Act (FIFRA). And, when consumers did read the labels, they thought that labels on household cleaning products not regulated by FIFRA were easier to read and understand than those on FIFRA-regulated products. Consumers in the EPA's focus groups also wanted the mandated labels to use fewer technical words, even though "public interest groups" recommended that full chemical names and other technical terms be provided. Consumers appear to have more common sense than their "consumer advocates."

As Judge Stephen Breyer noted in another context: "Who now reads the warnings on aspirin bottles, or the pharmaceutical drug warnings that run on, in tiny print, for several pages? Will a public that hears these warnings too often and too loudly begin to often to ignore them?" (Stephen Breyer, *Breaking the Vicious Circle*, Cambridge: Harvard University Press, 1993, p. 28). There is thus a downside to mandated information overload — there is a risk.

Negative Terms for Innovative Processes

There is also a risk that mandatory label information, especially for innovative food products, will be perceived as negative, especially when technical, little understood terms are used. The prevalence of negative terms for modern technological processes does not bode well for value-free mandated informational labeling of food produced by those processes. Rather, the words themselves to depict the processes would serve as warnings instead. The most prominent example of this would be foods produced by biotechnology.

Biotechnology is a modern technique that identifies and uses specific genes to modify plants, animals, and other organisms. As used in agriculture, biotechnology, because it is gene-specific, is a more precise technique than the conventional methods of cross-breeding and hybridization. Those older approaches, used almost since mankind began to cultivate crops and animals for food, attempted to enhance a desired trait or de-emphasize an undesirable one by cross-breeding. In early days, breeding techniques used wild species and bred them into cultivated species that had more desirable characteristics, for example, corn that grew larger or more evenly on the cob. Conventional cross-breeding can be inexact and may take lengthy periods of time for trial and error until a new plant exhibits the desired traits.

Modern biotechnology techniques, on the other hand, are targeted and fast, as scientists can insert specific genes that carry specific traits from one species to another.

In food production, biotechnology is currently being used to produce higher crop yields, by transferring genetic traits such as disease-resistance or pest-resistance into the plant itself. Crops are being bred to better resist droughts or floods or to grown in acidic or naturally toxic soils. The technique is also being used to enhance the nutrient levels of foods that are staples in many parts of the world, for instance, rice that is enhanced to produce high levels of Vitamin A, which could be used to correct a serious nutritional deficiency affecting millions of people and leading to blindness.

The FDA and Food Labeling

Currently the U.S. Food and Drug Administration (FDA) has policy guidelines about food product labeling. Those guidelines apply to foods and food ingredients, including those produced through use of biotechnology. New foods that contain a new substance or an allergen that is new to that food, or exhibit a different level of certain dietary nutrients or increased toxins are required to be tested, sometimes extensively, before being marketed. If a product in one of those categories is approved by the FDA, that food would also have to be labeled with information about the food's content and characteristics. However, if a peach is a peach is a peach — whether produced through conventional breeding techniques or through biotechnology, no special label is required.

It is interesting to note that when consumers have been polled about whether genetically modified food should be labeled, a majority says "yes." A consumer survey by the International Food Information Council (IFIC) also produced that result. However, when the IFIC survey explained the FDA's policy guidelines on labeling, most consumers thought that those guidelines applied to foods produced through biotechnology made sense.

A Set-Back for Biotech's Benefits

Critical public health benefits that biotechnology can offer could be set back rather than advanced through a restrictive approach to labeling information. The human and environmental benefits of agricultural biotechnology could be dramatic and widespread. Higher crop yields per acre can not only provide larger food output to feed the world's hungry, but also mean that less land would have to be used for farming, thus helping to preserve forests. The possible reduced use of pesticides can enhance the environment. The ability to grow crops in previously barren areas can help keep pace with the needs of growing populations, especially in developing countries. Enhanced nutritional levels of staple crops can prevent diseases that are life-threatening or debilitating.

Currently, consumers know little about those benefits. Activist groups that are campaigning against this new technology in countries across the world ignore the positive and instead promote images of fear and dread to depict the products of biotechnology. Some, such as Great Britain's Prince Charles, call biotech unethical and immoral. In parts of Europe, particularly in the United Kingdom, several radical groups have been burning test fields of genetically modified crops in protest. The small farmers affected by this destruction don't think those lawless gestures should be tolerated. They see themselves as part of the future of farming.

The anti-technology activists have been particularly vocal on the need for mandatory labeling of food produced through biotechnology. Under the mantra of "consumer choice" and consumers' "right to know," they want governments to mandate that all foods even with traces of genetically modified ingredients be labeled. Yet, it is more a matter of "some consumers wanting to know." Process information, such as how a food is produced, does not add any essential information for consumers. It may be something some people want to know — some consumers may be curious about learning about the process — but satisfying curiosity at the expense of critical information should not be the purpose of mandated labeling. That, however, is the type of situation that private markets can better address by providing truthful information about biotechnology's benefits through producers' claims on labels and through advertising.

"Organic" consumers and "organic" farmers, such as Prince Charles, are among some of the groups that are pushing for strict labeling of foods produced by biotechnology. It is ironic that the organic movement, in marketing and promoting their foodstuff, show that private markets provide consumers with information relating to values and choices. Organic producers often hawk their products as being "produced without genetic engineering." Thus, consumers who want such foods can readily locate them. Some consumers, many non-Jewish, are also turning to kosher foods because they want alternative foods and look to the kosher certification as a sign of quality.

Activists' push to label genetically modified foods would extend to those with even a trace of gene-spliced material. That approach will lead policy makers down a road to ridiculousness. In Germany, for example, regulators are pondering and meditating about whether a product that is washed in a bowl in which genetically modified food was put previously would have to be labeled. Where will that sort of approach lead — to witches reading entrails?

Over the last several decades, government has played an increasingly larger role in determining what information should be displayed on labels. To date, those mandates have been restricted to information of a specific, objective nature. It would clearly be inappropriate for government agencies to make judgments on a product's adherence to a religious code; it would also be inappropriate to mandate label information that is based on tastes or values. This distinction between objective and value-based information is critical. Now, however, some groups would claim for their values a special privilege. Groups opposed to biotechnology, food irradiation, and other modern food technologies seek special privileges and demand that their preferences should be observed on all products.

When government decides to mandate food labeling that relates to people's curiosity, their values, their wants, and their perceptions, it is taking on a role that will lead to more, rather than less misinformation, a "crowding out" of essential information, information overload, and warnings without cause. Proposals to require labeling of foods produced through biotechnology would likely cause some consumers to be misled into thinking that the food is less safe. Mandatory labeling that may be perceived as unnecessarily alarming can stand in the way of consumer acceptance of this process that could be invaluable in improving the world's food supply.

REASON * April 2000

Billions Served

Three decades after he launched the Green Revolution, agronomist Norman Borlaug is still fighting world hunger--and the doomsayers who say it's a lost cause.

Interviewed by Ronald Bailey

Who has saved more human lives than anyone else in history? Who won the Nobel Peace Prize in 1970? Who still teaches at Texas A&M at the age of 86? The answer is Norman Borlaug.

Who? Norman Borlaug, the father of the "Green Revolution," the dramatic improvement in agricultural productivity that swept the globe in the 1960s.

Borlaug grew up on a small farm in Iowa and graduated from the University of Minnesota, where he studied forestry and plant pathology, in the 1930s. In 1944, the Rockefeller Foundation invited him to work on a project to boost wheat production in Mexico. At the time Mexico was importing a good share of its grain. Borlaug and his staff in Mexico spent nearly 20 years breeding the high-yield dwarf wheat that sparked the Green Revolution, the transformation that forestalled the mass starvation predicted by neo-Malthusians.

In the late 1960s, most experts were speaking of imminent global famines in which billions would perish. "The battle to feed all of humanity is over," biologist Paul Ehrlich famously wrote in his 1968 bestseller The Population Bomb. "In the 1970s and 1980s hundreds of millions of people will starve to death in spite of any crash programs embarked upon now." Ehrlich also said, "I have yet to meet anyone familiar with the situation who thinks India will be self-sufficient in food by 1971." He insisted that "India couldn't possibly feed two hundred million more people by 1980."

But Borlaug and his team were already engaged in the kind of crash program that Ehrlich declared wouldn't work. Their dwarf wheat varieties resisted a wide spectrum of plant pests and diseases and produced two to three times more grain than the traditional varieties. In 1965, they had begun a massive campaign to ship the miracle wheat to Pakistan and India and teach local farmers how to cultivate it properly. By 1968, when Ehrlich's book appeared, the U.S. Agency for International Development had already hailed Borlaug's achievement as a "Green Revolution."

In Pakistan, wheat yields rose from 4.6 million tons in 1965 to 8.4 million in 1970. In India, they rose from 12.3 million tons to 20 million. And the yields continue to increase. Last year, India harvested a record 73.5 million tons of wheat, up 11.5 percent from 1998. Since Ehrlich's dire predictions in 1968, India's population has more than doubled, its wheat production has more than tripled, and its economy has grown nine-fold. Soon after Borlaug's success with wheat, his colleagues at the Consultative Group on International Agricultural Research developed high-yield rice varieties that quickly spread the Green Revolution through most of Asia.

Contrary to Ehrlich's bold pronouncements, hundreds of millions didn't die in massive famines. India fed far more than 200 million more people, and it was close enough to self-sufficiency in food production by 1971 that Ehrlich discreetly omitted his prediction about that from later editions of The Population Bomb. The last four decades have seen a "progress explosion" that has handily outmatched any "population explosion."

Borlaug, who unfortunately is far less well-known than doom-sayer Ehrlich, is responsible for much of the progress humanity has made against hunger. Despite occasional local famines caused by armed conflicts or political mischief, food is more abundant and cheaper today than ever before in history, due in large part to the work of Borlaug and his colleagues.

More than 30 years ago, Borlaug wrote, "One of the greatest threats to mankind today is that the world may be choked by an explosively pervading but well camouflaged bureaucracy." As REASON's interview with him shows, he still believes that environmental activists and their allies in international agencies are a threat to progress on global food security. Barring such interference, he is confident that agricultural research, including biotechnology, will be able to boost crop production to meet the demand for food in a world of 8 billion or so, the projected population in 2025.

Meanwhile, media darlings like Worldwatch Institute founder Lester Brown keep up their drumbeat of doom. In 1981 Brown declared, "The period of global food security is over." In 1994, he wrote, "The world's farmers can no longer be counted on to feed the projected additions to our numbers." And as recently as 1997 he warned, "Food scarcity will be the defining issue of the new era now unfolding, much as ideological conflict was the defining issue of the historical era that recently ended."

Borlaug, by contrast, does not just wring his hands. He still works to get modern agricultural technology into the hands of hungry farmers in the developing world. Today, he is a consultant to the International Maize and Wheat Center in Mexico and president of the Sasakawa Africa Association, a private Japanese foundation working to spread the Green Revolution to sub-Saharan Africa.

REASON Science Correspondent Ronald Bailey met with Borlaug at Texas A&M, where he is Distinguished Professor in the Soil and Crop Sciences Department and still teaches classes on occasion. Despite his achievements, Borlaug is a modest man who works out of a small windowless office in the university's agricultural complex. A few weeks before the interview, Texas A&M honored Borlaug by naming its new agricultural biotechnology center after him. "We have to have this new technology if we are to meet the growing food needs for the next 25 years," Borlaug declared at the dedication ceremony. If the naysayers do manage to stop agricultural biotech, he fears, they may finally bring on the famines they have been predicting for so long.

Reason: What are you currently working on?

Norman Borlaug: Since 1984, I've been involved in the Sasakawa Africa Association. Our program has devised the best package of farming practices we could with the best seed available, the best agronomic practices, the best rates and dates of seeding, the best controls for weeds and insects and diseases, and put them into test plots in 14 countries. We have found that there is a large food production potential in these African countries which are now struggling with food shortages. The package of practices that we have devised uses modest levels of inputs so the cost is not particularly high compared to their traditional ways of farming. The yields are at the worst double, nearly always triple, and sometimes quadruple what the traditional practices are producing. African farmers are very enthusiastic about these new methods.

Reason: Could genetically engineered crops help farmers in developing countries?

Borlaug: Biotech has a big potential in Africa, not immediately, but down the road. Five to eight years from now, parts of it will play a role there. Take the case of maize with the gene that controls the tolerance level for the weed killer Roundup. Roundup kills all the weeds, but it's short-lived, so it doesn't have any residual effect, and from that standpoint it's safe for people and the environment. The gene for herbicide tolerance is built into the crop variety, so that when a farmer sprays he kills only weeds but not the crops. Roundup Ready soybeans and corn are being very widely used in the U.S. and Argentina. At this stage, we haven't used varieties with the tolerance for Roundup or any other weed killer [in Africa], but it will have a role to play.

Roundup Ready crops could be used in zero-tillage cultivation in African countries. In zero tillage, you leave the straw, the rice, the wheat if it's at high elevation, or most of the corn stock, remove only what's needed for animal feed, and plant directly [without plowing], because this will cut down erosion. Central African farmers don't have any animal power, because sleeping sickness kills all the animals--cattle, the horses, the burros and the mules. So draft animals don't exist, and farming is all by hand and the hand tools are hoes and machetes. Such hand tools are not very effective against the aggressive tropical grasses that typically invade farm fields. Some of those grasses have sharp spines on them, and they're not very edible. They invade the cornfields, and it gets so bad that farmers must abandon the fields for a while, move on, and clear some more forest. That's the way it's been going on for centuries, slash-and-burn farming. But with this kind of weed killer, Roundup, you can clear the fields of these invasive grasses and plant directly if you have the herbicide-tolerance gene in the crop plants.

Reason: What other problems do you see in Africa?

Borlaug: Supplying food to sub-Saharan African countries is made very complex because of a lack of infrastructure. For example, you bring fertilizer into a country like Ethiopia, and the cost of transporting the fertilizer up the mountain a few hundred miles to Addis Ababa doubles its cost. All through sub-Saharan Africa, the lack of roads is one of the biggest obstacles to development--and not just from the standpoint of moving agricultural inputs in and moving increased grain production to the cities. That's part of it, but I think roads also have great indirect value. If a road is built going across tribal groups and some beat-up old bus starts moving, in seven or eight years you'll hear people say, "You know, that tribe over there, they aren't so different from us after all, are they?"

And once there's a road and some vehicles moving along it, then you can build schools near a road. You go into the bush and you can get parents to build a school from local materials, but you can't get a teacher to come in because she or he will say, "Look, I spent six, eight years preparing myself to be a teacher. Now you want me to go back there in the bush? I won't be able to come out and see my family or friends for eight, nine months. No, I'm not going." The lack of roads in Africa greatly hinders agriculture, education, and development.

Reason: Environmental activists often oppose road building. They say such roads will lead to the destruction of the rain forests or other wildernesses. What would you say to them?

Borlaug: These extremists who are living in great affluence...are saying that poor people shouldn't have roads. I would like to see them not just go out in the bush backpacking for a week but be forced to spend the rest of their lives out there and have their children raised out there. Let's see whether they'd have the same point of view then.

I should point out that I was originally trained as a forester. I worked for the U.S. Forest Service, and during one of my assignments I was reputed to be the most isolated member of the Forest Service, back in the middle fork of the Salmon River, the biggest primitive area in the southern 48 states. I like the back country, wildlife and all of that, but it's wrong to force poor people to live that way.

Reason: Does the European ban on biotechnology encourage elites in developing countries to say, "Well, if it's not good enough for Europeans, it's not good enough for my people"?

Borlaug: Of course. This is a negative effect. We always have this. Take the case of DDT. When it was banned here in the U.S. and the European countries, I testified about the value of DDT for malaria control, especially throughout Africa and in many parts of Asia. The point I made in my testimony as a witness for the USDA was that if you ban DDT here in the U.S., where you don't have these problems, then OK, you've got other insecticides for agriculture, but when you ban it here and then exert pressures on heads of government in Africa and Asia, that's another matter. They've got serious human and animal diseases, and DDT is important. Of course, they did ban DDT, and the danger is that they will do the same thing with biotech now.

Reason: What do you see as the future of biotechnology in agriculture?

Borlaug: Biotechnology will help us do things that we couldn't do before, and do it in a more precise and safe way. Biotechnology will allow us to cross genetic barriers that we were never able to cross with conventional genetics and plant breeding. In the past, conventional plant breeders were forced to bring along many other genes with the genes, say, for insect or disease resistance that we wanted to incorporate in a new crop variety. These extra genes often had negative effects, and it took years of breeding to remove them. Conventional plant breeding is crude in comparison to the methods that are being used with genetic engineering. However, I believe that we have done a poor job of explaining the complexities and the importance of biotechnology to the general public.

Reason: A lot of activists say that it's wrong to cross genetic barriers between species. Do you agree?

Borlaug: No. As a matter of fact, Mother Nature has crossed species barriers, and sometimes nature crosses barriers between genera--that is, between unrelated groups of species. Take the case of wheat. It is the result of a natural cross made by Mother

Nature long before there was scientific man. Today's modern red wheat variety is made up of three groups of seven chromosomes, and each of those three groups of seven chromosomes came from a different wild grass. First, Mother Nature crossed two of the grasses, and this cross became the durum wheats, which were the commercial grains of the first civilizations spanning from Sumeria until well into the Roman period. Then Mother Nature crossed that 14-chromosome durum wheat with another wild wheat grass to create what was essentially modern wheat at the time of the Roman Empire.

Durum wheat was OK for making flat Arab bread, but it didn't have elastic gluten. The thing that makes modern wheat different from all of the other cereals is that it has two proteins that give it the doughy quality when it's mixed with water. Durum wheats don't have gluten, and that's why we use them to make spaghetti today. The second cross of durum wheat with the other wild wheat produced a wheat whose dough could be fermented with yeast to produce a big loaf. So modern bread wheat is the result of crossing three species barriers, a kind of natural genetic engineering.

Reason: Environmentalists say agricultural biotech will harm biodiversity.

Borlaug: I don't believe that. If we grow our food and fiber on the land best suited to farming with the technology that we have and what's coming, including proper use of genetic engineering and biotechnology, we will leave untouched vast tracts of land, with all of their plant and animal diversity. It is because we use farmland so effectively now that President Clinton was recently able to set aside another 50 or 60 million acres of land as wilderness areas. That would not have been possible had it not been for the efficiency of modern agriculture.

In 1960, the production of the 17 most important food, feed, and fiber crops--virtually all of the important crops grown in the U.S. at that time and still grown today--was 252 million tons. By 1990, it had more than doubled, to 596 million tons, and was produced on 25 million fewer acres than were cultivated in 1960. If we had tried to produce the harvest of 1990 with the technology of 1960, we would have had to have increased the cultivated area by another 177 million hectares, about 460 million more acres of land of the same quality--which we didn't have, and so it would have been much more. We would have moved into marginal grazing areas and plowed up things that wouldn't be productive in the long run. We would have had to move into rolling mountainous country and chop down our forests. President Clinton would not have had the nice job of setting aside millions of acres of land for restricted use, where you can't cut a tree even for paper and pulp or for lumber. So all of this ties together.

This applies to forestry, too. I'm pleased to see that some of the forestry companies are very modern and using good management, good breeding systems. Weyerhauser is Exhibit A. They are producing more wood products per unit of area than the old unmanaged forests. Producing trees this way means millions of acres can be left to natural forests.

Reason: A lot of environmental activists claim that the BT toxin gene, which is derived from *Bacillus thuringiensis* and which has been transferred into corn and cotton, is going to harm beneficial insects like the monarch butterfly. Is there any evidence of that?

Borlaug: To that I [respond], will BT harm beneficial insects more than the insecticides that are sprayed around in big doses? In fact, BT is more specific. There are lots of insects that it doesn't affect at all.

Reason: It affects only the ones that eat the crops.

Borlaug: Right.

Reason: So you don't think that putting the BT gene in corn or cotton is a big problem?

Borlaug: I think that whole monarch butterfly thing was a gross exaggeration. I think the researchers at Cornell who fed BT corn pollen to monarch butterflies were looking for something that would make them famous and create this big hullabaloo that's resulted. In the first place, corn pollen is pretty heavy. It doesn't fly long distances. Also, most monarchs are moving at different times of the season when there's no corn pollen. Sure, some of them might get killed by BT corn pollen, but how many get killed when they are sprayed with insecticides? Activists also say that BT genes in crops will put stress on the pest insects, and they'll mutate. Well, that's been going on with conventional insecticides. It's been going on all my life working with wheat. It's a problem that has been and can be managed.

Reason: But the Cornell researchers went ahead and published their paper on the effects of BT corn pollen on monarch butterflies in the laboratory.

Borlaug: Several of us tried to encourage them to run field tests before it was published. That's how science gets politicized. There's an element of Lysenkoism [Lysenko was Stalin's favorite biologist] all tangled up with this pseudoscience and environmentalism. I like to remind my friends what pseudoscience and misinformation can do to destroy a nation.

Reason: Some activists claim that herbicide-resistant crops end up increasing the amount of herbicide that's sprayed on fields. Do you think that's true?

Borlaug: Look, insecticides, herbicides, and fertilizer cost money, and the farmer doesn't have much margin. He's going to try to use the minimum amount that he can get by with. Probably in most cases, a farmer applies less than he should. I don't think farmers are likely to use too much.

Reason: What other crop pests might biotech control in the future?

Borlaug: All of the cereals except rice are susceptible to one to three different species of rust fungi. Now, rusts are obligate parasites. They can only live under green tissue, but they are long-lived. They can move in the air sometimes 100, 500, 800 miles, and they get in the jet stream and fall. If the crop variety is susceptible to rust fungi and moisture is there and the temperature is right, it's like lighting a fire. It just destroys crops. But rice isn't susceptible--no rust....One thing that I hope to live to see is somebody taking that block of rust-resistance genes in rice and putting it into all of the other cereals.

Reason: Do biotech crops pose a health risk to human beings?

Borlaug: I see no difference between the varieties carrying a BT gene or a herbicide resistance gene, or other genes that will come to be incorporated, and the varieties created by conventional plant breeding. I think the activists have blown the health risks of biotech all out of proportion.

Reason: What do you think of organic farming? A lot of people claim it's better for human health and the environment.

Borlaug: That's ridiculous. This shouldn't even be a debate. Even if you could use all the organic material that you have--the animal manures, the human waste, the plant residues--and get them back on the soil, you couldn't feed more than 4 billion people. In addition, if all agriculture were organic, you would have to increase cropland area dramatically, spreading out into marginal areas and cutting down millions of acres of forests.

At the present time, approximately 80 million tons of nitrogen nutrients are utilized each year. If you tried to produce this nitrogen organically, you would require an additional 5 or 6 billion head of cattle to supply the manure. How much wild land would you have to sacrifice just to produce the forage for these cows? There's a lot of nonsense going on here.

If people want to believe that the organic food has better nutritive value, it's up to them to make that foolish decision. But there's absolutely no research that shows that organic foods provide better nutrition. As far as plants are concerned, they can't tell whether that nitrate ion comes from artificial chemicals or from decomposed organic matter. If some consumers believe that it's better from the point of view of their health to have organic food, God bless them. Let them buy it. Let them pay a bit more. It's a free society. But don't tell the world that we can feed the present population without chemical fertilizer. That's when this misinformation becomes destructive.

Reason: What do you think of Worldwatch Institute founder Lester Brown and his work?

Borlaug: I've known Lester Brown personally for more than 40 years. He's done a lot of good, but he vacillates, depending on the way the political and economic winds are blowing, and he's sort of inclined to be a doomsayer.

Reason: He recently said, "The world's farmers can no longer be counted on to feed the projected additions to our numbers." Do you agree with that?

Borlaug: No, I do not. With the technology that we now have available, and with the research information that's in the pipeline and in the process of being finalized to move into production, we have the know-how to produce the food that will be needed to feed the population of 8.3 billion people that will exist in the world in 2025.

I don't like to try to see further than about 25 years. In 1970, at the Nobel Prize press conference, I said I can see that we have the technology to produce the food that's needed to the year 2000, and that we can do it without destroying a lot of the environment. Modern agriculture saves a lot of land for nature, for wildlife habitat, for flood control, for erosion control, for forest production. All of those are values that are important to society in general, and especially to the privileged who have a chance to spend a lot of long vacations out looking at nature. I say we can produce enough food with the technology available and what's in the process of being developed, assuming that we don't have all this agricultural progress destroyed by the doomsayers. That is, we will be able to produce enough food in 2025 without expanding the area under cultivation very much and without having to move into semi-arid or forested mountainous topographies.

Reason: It seems that every five years or so, Lester Brown predicts that massive famines are imminent. Why does he do that? They never happen.

Borlaug: I guess it sells. I guess what he writes has a lot to do with raising funds.

Reason: Brown notes that India tripled its wheat yields in the past three decades, but he says that will be impossible to do again. Do you think he's right?

Borlaug: No. The projections in food production in India continue to go up on the same slope. When we transferred the Green Revolution wheat technology to India, production was 12 million tons a year. Last year it was 74 million tons, and it is still going up. Once in a while production may go down by a couple of million tons when there's a drought, but in general it continues to go up. Also, the increase in production has occurred with very modest increases in cultivated area. A lot of wild land has been saved in India, China, and the United States by high-yield technology.

India has produced enough and sometimes has a surplus in grain. The problem is to get it into the stomachs of the hungry. There's a lack of purchasing power by too large a part of the population. There are still many hungry people, not dying from starvation, but needing more food to grow strong bodies and maintain health and work effectively. The grain is there in the warehouses, but it doesn't find its way into the stomachs of the hungry.

Reason: What do you think of Paul Ehrlich's work?

Borlaug: Ehrlich has made a great career as a predictor of doom. When we were moving the new wheat technology to India and Pakistan, he was one of the worst critics we had. He said, "This person, Borlaug, doesn't have any idea of the magnitude of the problems in food production." He said, "You aren't going to make any major impact on producing the food that's needed." Despite his criticisms, we succeeded, of course.

Reason: When an alleged expert like Ehrlich is being negative like that, does that discourage people? Does it hurt the efforts to boost food production?

Borlaug: Sure, because we were funded by a foundation....They'd hear his criticisms, and I'm sure there were some people at Rockefeller saying, "Maybe we shouldn't fund that program anymore." It always has adverse effects on budgeting.

Reason: Why do you think people still listen to Ehrlich? One can go back and read his doomsday scenarios and see that he was wrong.

Borlaug: People don't go back and read what he wrote. You do, but the great majority of the people don't, and their memory is short. As a matter of fact, I think this [lack of perspective] is true of our whole food situation. Our elites live in big cities and are far removed from the fields. Whether it's Brown or Ehrlich or the head of the Sierra Club or the head of Greenpeace, they've never been hungry.

Reason: You mentioned that you are afraid that the doomsayers could stop the progress in food production.

Borlaug: It worries me, if they gum up all of these developments. It's elitism, and the American people are vulnerable to this, too. I'm talking about the extremists here and in Western Europe....In the U.S., 98 percent of consumers live in cities or urban areas or good-size towns. Only 2 percent still live out there on the land. In Western Europe also, a big percentage of the people live off the farms, and they don't understand the complexities of agriculture. So they are easily swayed by these scare stories that we are on the verge of being poisoned out of existence by farm chemicals.

Bruce Ames, the head of biochemistry at Berkeley, has analyzed hundreds and hundreds of foods, including all of the basic ones that we have been eating from the beginning of agriculture up to the present time. He has found that they contain trace amounts of many completely natural chemical compounds that are toxic or carcinogenic, but they're present in such small quantities that they apparently don't affect us.

Reason: Would you say the Green Revolution was a success?

Borlaug: Yes, but it's a never-ending job. When I was born in 1914, the world population was approximately 1.6 billion people. It has just turned 6 billion. We've had no major famines any place in the world since the Green Revolution began. We've had local famines where these African wars have been going on and are still going on. However, if we could get the infrastructure straightened out in African countries south of the Sahara, you could end hunger there pretty fast....And if you look at the data that's put out by the World Health Organization and [the U.N.'s Food and Agriculture Organization], there are probably 800 million people who are undernourished in the world. So there's still a lot of work to do.

The New Monsanto Pledge

In 1990, Monsanto announced the Monsanto Pledge — a statement of environmental responsibility. We achieved virtually all of what we pledged to do.

But one principle for which we still have work to do applies directly to the issue of biotechnology: That principle states simply, "We will work to achieve sustainable agriculture through new technology and new practices."

Today Monsanto is a new company, solely devoted to agriculture. And we have developed a new pledge, to help us fulfill our promise for sustainable agriculture. This new Monsanto pledge includes the following five elements — dialogue, transparency, respect, sharing and delivering benefits.

Dialogue

- We commit to an ongoing dialogue with all interested parties to understand the issues and concerns related to this technology.
  - To this end, we commit to create an external Biotechnology Advisory Council from a range of constituencies with an interest in biotechnology to meet, discuss, advise and help us make decisions.
  - And we commit to involving our customers to help us make decisions about the development, use and stewardship of new agricultural technologies.

Transparency

- We commit to transparency by making published scientific data and data summaries on product safety and benefits publicly available and accessible, and we commit to working within the rigorous, science-based regulation as required by appropriate government agencies around the world.
  - We will make both Monsanto research and external research by universities and other institutions available through the Internet and other public venues.
  - We commit our support for a mandatory pre-market notification process for Food and Drug Administration (FDA) review of all biotechnology products in the United States.
  - We commit to work toward the establishment of global standards for the quality of seed, grain and food products.

Respect

- We commit to respecting the religious, cultural and ethical concerns of people throughout the world by:
  - Commercializing commodity grain products only after they have been approved for consumption by both humans and animals;
  - Not using genes taken from animal or human sources in our agricultural products intended for food or feed;
  - Never commercializing a product in which a known allergen has been introduced;
  - Using alternatives to antibiotic resistance genes to select for new traits as soon as the technology allows us to do so efficiently and effectively in a manner that has been proven safe; and,
  - Underscoring our commitment not to pursue technologies that result in sterile seeds.

Sharing

- We commit to bring the knowledge and advantages of all forms of agriculture to resource-poor farmers in the developing world to help improve food security and protect the environment.
  - To this end, we have created a dedicated team within Monsanto to facilitate technology sharing and agricultural development collaborations with public institutions, non-profit groups and local industry around the world.

Benefits

- We commit to work for and deliver benefits for farmers commercially as well as environmentally.
  - Environmentally, we commit to develop technology that directly contributes to a vision of abundant food and a healthy environment by:
    - Using biotechnology to promote integrated pest management (IPM) and reduce agricultural inputs, such as we have seen with the reduction of pesticides in the United States;
    - Working with growers worldwide to double the use of tillage practices that conserve soil and moisture over the next five years; and
    - Ensuring that all of our products and practices protect wildlife and beneficial species.
  - Commercially, we intend to launch new genetically improved commodity crops in the United States only after they have received full approval for food use and animal feed in the United States and Japan. We hope also to extend this intention to

Europe as soon as it has established a working regulatory system. We're able to state this intention as long as there are science-based regulatory systems that make timely decisions. If the regulatory systems are not functional, we cannot allow the breakdown to deny U.S. farmers the choice of new technologies.

---

http://www.monsanto.com/monsanto/about_us/monsanto_pledge/default.htm

In All Fairness

21st Century Luddites?

Stopping technology that can feed the world

I never thought I'd see the day when the same crowd who urged us to "feed the world" during the famines of the 1980s would attack technological advances that could eliminate world hunger. But this is exactly what is happening now, as professional activists engineer a campaign to obstruct the scientific enhancement of our food supply.

This scientific process — the genetic modification and enhancement of foods — is simply a modern version of age-old techniques used by farmers to create produce like seedless grapes and nectarines. The new technologies are even safer and more efficient. Thousands of tests performed worldwide support American regulators' findings that enhanced crops pose no special risk for human health or the environment. Naysaying activists can't credibly point to a single person, or even one mouse, that has fallen ill or died from consuming genetically-modified foods.

The special interests refuse to recognize the enormous public health benefits these products have to offer. In addition to enhancing foods' taste and shelf life, bioengineering enables more crops to grown on less land in even the harshest of climates. Around the world, over 170 million preschool children suffer from malnutrition. Biotechnology, if allowed to progress, could dramatically improve their health and alter the lives of the millions who now go to bed hungry. Testing has shown that by genetically modifying the iron and vitamin A content in rice, we can reduce childhood anemia and blindness.

In addition, genetic crop enhancement can also help environmental groups achieve such long-standing goals as reduction of pesticide use and ground-water pollution, control of topsoil erosion, and the preservation of wilderness.

But instead of helping to guide these developments forward, the special interests have executed a smear campaign straight out of the Activism 101 textbook. Some radicals have even sabotaged genetic crop testing by destroying acres of modified plants.

It is puzzling why, when the genius of science offers such remarkable benefits to humanity at little or no risk, the professional activists would go to such lengths to stop it. Has their visceral bias against free enterprise clouded their judgment, or are they embracing radical Luddites' disdain for technology? Either way, their actions are a slap at philanthropies and apolitical scientists who have long fought hunger and starvation.

Technology has given us an opportunity to dramatically expand and improve the world's food supply *and* protect the environment. Unfortunately, myopic special interests have a very different vision for the future, one which they seemingly want to achieve by bringing progress to a screeching halt. Demonizing the biotechnology bogeyman may keep financial contributions rolling in and put food on well-fed activists' tables, but it does little to alleviate world hunger. Maybe when the public realizes this, they will finally lose their appetite for the antics of self-serving "public interest" groups.

Source: The New York Times Op-Ed, Monday, January 24, 2000

WLF's Civic Communications Program publishes "In All Fairness" in the national edition of the *New York Times.* The op-ed feature reaches 3.2 million readers in 70 major markets, as well as a diverse group of thought leaders, decision makers, and the public.

Section VII

International Documents

The role that international concerns, beliefs, and cultures have played in raising the political profile of genetically modified food export, distribution, and consumption pervades to some degree every section of this *Casebook,* and cannot simply be broken out of its broad context. But this section will address some formal actions that are noteworthy. In general, the complexity of issues related to the production, distribution, and sale of genetically modified foods are a rare example of a situation that the U.S. finds itself unable to control, despite its overwhelming influence on world politics generally and its enormous economic power. It is also unusual for the U.S. to be favoring reduced regulation rather than, as the *Environmental Politics: Interest Groups, the Media, and the Making of Policy* documents, having its generally more stringent levels of regulation eroded by the WTO and other international bodies.

The two documents, reproduced here, will significantly shape the way that European anxieties will be reconciled with U.S. exports. In time, we may look upon them as pivotal elements in forging international policy on genetically modified foods. The first is a regulation adopted by the European Community (EU) in May of 1988. Specifically, it requires the labeling of genetically modified soy and corn prior to delivery to the consumer, and it prescribes the form of that label in Article 2. The regulation is binding on all members of the EU. More interesting are the "Whereas" clauses that precede the substantive provisions, which systematically and incrementally develop an argument justifying the regulation. The "Whereas" clauses are a counterpart to what in U.S. state and federal laws are identified as "Findings and Declarations" sections or, simply "Findings."

The second document is the *Cartagena Protocol on Biosafety to the Convention on Biological Diversity.* The Cartagena Protocol is an international agreement entered into by more than 130 nations, including the U.S., in January of 2000 and ratified later that spring, to govern trade in genetically engineered commodities, e.g., seeds, animals, microbes, and crops, but not pharmaceuticals. Specifically, it authorizes a member nation to ban imports of a genetically modified product if it believes that there is insufficient evidence that the product is safe. It also establishes rules governing their transport, requiring that the words "may contain modified living organisms" appear on all shipments of genetically modified foods.

Discussion

The potential implications of these two international efforts on U.S. policy regarding genetically modified food are enormous. The EU regulation requiring labeling will no doubt impact the debate over the aspect of the issue that has captured the most attention in this country. It would be profitable to explore the significance of labeling relative to other aspects of the issue, and why and how a foreign requirement will affect not only U.S. imports, but domestic sale and consumption. The discussion may well throw light on why herbal remedies here have not incurred the wrath of the environmental and health communities that genetically modified foods have, and why genetically modified foods constitute a more enticing media subject. It would also prove useful to compare the argument in the preamble of the regulation to those made by anti-genetically modified foods forces here.

Like most other treaties involving diverse interests, the Cartagena Protocol provides something for all sides to celebrate and gives competing countries grounds for claiming a measure of victory. Environmentalists are pleased with several elements of the treaty: 1) at least tacitly, it affirms that GMOs (here referred to as LMOs—"living modified organisms") are different in some essential way from traditionally bred organisms and are thus not the "substantial equivalent" of food produced by conventional breeding methods that the U.S. has used to justify their regulatory inaction; 2) it specifically establishes the precautionary principle in international law, allowing nations to invoke it to prohibit the importation of genetically modified foods if it believes that they would pose risks to human health or the environment; 3) it imposes the burden to establish the safety of GMOs in transport on the exporter, who must also provide advanced notice of such export; 4) it permits consideration of socio-economic considerations in their risk assessment; and 5) it specifically affirms that the Protocol is not subordinate to other international agreements (e.g., those fashioned by the WTO), though, on the other hand, it doesn't supersede others.

Supporters of freer trade policies with regard to GMOs are consoled that the treaty exempts pharmaceuticals and commodities not destined for direct release into the environment such as grains or cotton, and that it has no provisions for liability, which have been put off to a later day. Study of the "fuzzy" language regarding such hot button issues as scientific certainty and the nature of commodities that "may contain" genetic components also may go a long way toward explaining how this "compromise" measure managed to secure the formal signatures of such a large and diverse body of countries.

Individually or jointly, these documents provide fertile ground for exploring the political roots of the controversy, the impact that treatment of genetically modified foods abroad may have on U.S. policy, and speculating on how they may affect third world countries, to whom biotech interests are holding out the promise of nutrition that genetically modified foods may uniquely supply.

The shifting tides of the controversy may be followed almost daily on the Internet, but the sites of the biotech industry, www.whybiotech.com and www.monsanto.com, for example, as well as those most aggressively opposing them among the environmental advocacy groups, www.greenpeace.org and www.foe.org/safefood, for example, are a good start. Links from these will carry the site visitor into other areas.

Cartagena Protocol on Biosafety to the Convention on Biological Diversity

The Parties to this Protocol,

Being Parties to the Convention on Biological Diversity, hereinafter referred to as "the Convention",

Recalling Article 19, paragraphs 3 and 4, and Articles 8 (g) and 17 of the Convention,

Recalling also decision II/5 of 17 November 1995 of the Conference of the Parties to the Convention to develop a Protocol on biosafety, specifically focusing on transboundary movement of any living modified organism resulting from modern biotechnology that may have adverse effect on the conservation and sustainable use of biological diversity, setting out for consideration, in particular, appropriate procedures for advance informed agreement,

Reaffirming the precautionary approach contained in Principle 15 of the Rio Declaration on Environment and Development,

Aware of the rapid expansion of modern biotechnology and the growing public concern over its potential adverse effects on biological diversity, taking also into account risks to human health,

Recognizing that modern biotechnology has great potential for human well-being if developed and used with adequate safety measures for the environment and human health,

Recognizing also the crucial importance to humankind of centres of origin and centres of genetic diversity,

Taking into account the limited capabilities of many countries, particularly developing countries, to cope with the nature and scale of known and potential risks associated with living modified organisms,

Recognizing that trade and environment agreements should be mutually supportive with a view to achieving sustainable development,

Emphasizing that this Protocol shall not be interpreted as implying a change in the rights and obligations of a Party under any existing international agreements,

Understanding that the above recital is not intended to subordinate this Protocol to other international agreements,

Have agreed as follows:

Article 1
OBJECTIVE

In accordance with the precautionary approach contained in Principle 15 of the Rio Declaration on Environment and Development, the objective of this Protocol is to contribute to ensuring an adequate level of protection in the field of the safe transfer, handling and use of living modified organisms resulting from modern biotechnology that may have adverse effects on the conservation and sustainable use of biological diversity, taking also into account risks to human health, and specifically focusing on transboundary movements.

Article 2
GENERAL PROVISIONS

1. Each Party shall take necessary and appropriate legal, administrative and other measures to implement its obligations under this Protocol.
2. The Parties shall ensure that the development, handling, transport, use, transfer and release of any living modified organisms are undertaken in a manner that prevents or reduces the risks to biological diversity, taking also into account risks to human health.
3. Nothing in this Protocol shall affect in any way the sovereignty of States over their territorial sea established in accordance with international law, and the sovereign rights and the jurisdiction which States have in their exclusive economic zones and their continental shelves in accordance with international law, and the exercise by ships and aircraft of all States of navigational rights and freedoms as provided for in international law and as reflected in relevant international instruments.
4. Nothing in this Protocol shall be interpreted as restricting the right of a Party to take action that is more protective of the conservation and sustainable use of biological diversity than that called for in this Protocol, provided that such action is consistent with the objective and the provisions of this Protocol and is in accordance with that Party's other obligations under international law.
5. The Parties are encouraged to take into account, as appropriate, available expertise, instruments and work undertaken in international forums with competence in the area of risks to human health.

Article 3
USE OF TERMS

For the purpose of this Protocol:

(a) "Conference of the Parties" means the Conference of the Parties to the Convention;

(b) "Contained use" means any operation, undertaken within a facility, installation or other physical structure, which involves living modified organisms that are controlled by specific measures that effectively limit their contact with, and their impact on, the external environment;

(c) "Export" means intentional transboundary movement from one Party to another Party;

(d) "Exporter" means any legal or natural person, under the jurisdiction of the Party of export, who arranges for a living modified organism to be exported;

(e) "Import" means intentional transboundary movement into one Party from another Party;

(f) "Importer" means any legal or natural person, under the jurisdiction of the Party of import, who arranges for a living modified organism to be imported;

(g) "Living modified organism" means any living organism that possesses a novel combination of genetic material obtained through the use of modern biotechnology;

(h) "Living organism" means any biological entity capable of transferring or replicating genetic material, including sterile organisms, viruses and viroids;

(i) "Modern biotechnology" means the application of:

a. *In vitro* nucleic acid techniques, including recombinant deoxyribonucleic acid (DNA) and direct injection of nucleic acid into cells or organelles, or

b. Fusion of cells beyond the taxonomic family, that overcome natural physiological reproductive or recombination barriers and that are not techniques used in traditional breeding and selection;

(j) "Regional economic integration organization" means an organization constituted by sovereign States of a given region, to which its member States have transferred competence in respect of matters governed by this Protocol and which has been duly authorized, in accordance with its internal procedures, to sign, ratify, accept, approve or accede to it;

(k) "Transboundary movement" means the movement of a living modified organism from one Party to another Party, save that for the purposes of Articles 17 and 24 transboundary movement extends to movement between Parties and non-Parties.

Article 4
SCOPE

This Protocol shall apply to the transboundary movement, transit, handling and use of all living modified organisms that may have adverse effects on the conservation and sustainable use of biological diversity, taking also into account risks to human health.

Article 5
PHARMACEUTICALS

Notwithstanding Article 4 and without prejudice to any right of a Party to subject all living modified organisms to risk assessment prior to the making of decisions on import, this Protocol shall not apply to the transboundary movement of living modified organisms which are pharmaceuticals for humans that are addressed by other relevant international agreements or organizations.

Article 6
TRANSIT AND CONTAINED USE

1. Notwithstanding Article 4 and without prejudice to any right of a Party of transit to regulate the transport of living modified organisms through its territory and make available to the Biosafety Clearing-House, any decision of that Party, subject to Article 2, paragraph 3, regarding the transit through its territory of a specific living modified organism, the provisions of this Protocol with respect to the advance informed agreement procedure shall not apply to living modified organisms in transit.
2. Notwithstanding Article 4 and without prejudice to any right of a Party to subject all living modified organisms to risk assessment prior to decisions on import and to set standards for contained use within its jurisdiction, the provision of this Protocol with respect to the advance informed agreement procedure shall not apply to the transboundary movement of living modified organisms destined for contained use undertaken in accordance with the standards of the Party of import.

Article 7
APPLICTON OF THE ADVANCE INFORMED AGREEMENT PROCEDURE

1. Subject to Articles 5 and 6, the advance informed agreement procedure in Articles 8 to 10 and 12 shall apply prior to the first intentional transboundary movement of living modified organisms for intentional introduction into the environment of the Party of import.
2. "Intentional introduction into the environment" in paragraph 1 above, does not refer to living modified organisms intended for direct use as food or feed, or for processing.
3. Article 11 shall apply prior to the first transboundary movement of living modified organisms intended for direct use as food or feed, or for processing.
4. The advance informed agreement procedure shall not apply to the intentional transboundary movement of living modified organisms identified in a decision of the Conference of the Parties serving as a meeting of the Parties to this Protocol as being not likely to have adverse effects on the conservation and sustainable use of biological diversity, taking also into account risks to human health.

Article 8
NOTIFICATION

1. The Party of export shall notify, or require the exporter to ensure notification to, in writing, the competent national authority of the Party of import prior to the intentional transboundary movement of a living modified organism that falls within the scope of Article 7, paragraph 1. The notification shall contain, at a minimum, the information specified in Annex I.
2. The Party of export shall ensure that there is a legal requirement for the accuracy of information provided by the exporter.

Article 9
ACKNOWLEDGEMENT OF RECEIPT OF NOTIFICATION

1. The Party of import shall acknowledge receipt of the notification, in writing, to the notifier within ninety days of its receipt.
2. The acknowledgement shall state:
   (a) The date of receipt of the notification;
   (b) Whether the notification, prima facie, contains the information referred to in Article 8;
   (c) Whether to proceed according to the domestic regulatory framework of the Party of import or according to the procedure specified in Article 10.
3. The domestic regulatory framework referred to in paragraph 2 (c) above, shall be consistent with this Protocol.
4. A failure by the Party of import to acknowledge receipt of a notification shall not imply its consent to an intentional transboundary movement.

Article 10
DECISION PROCEDURE

1. Decisions taken by the Party of import shall be in accordance with Article 15.
2. The Party of import shall, within the period of time referred to in Article 9, inform the notifier, in writing, whether the intentional transboundary movement may proceed:
   (a) Only after the Party of import has given its written consent; or
   (b) After no less than ninety days without a subsequent written consent.
3. Within two hundred and seventy days of the date of receipt of notification, the Party of import shall communicate, in writing, to the notifier and to the Biosafety Clearing-House the decision referred to in paragraph 2 (a) above:

(a) Approving the import, with or without conditions, including how the decision will apply to subsequent imports of the same living modified organism;
(b) Prohibiting the import;
(c) Requesting additional relevant information in accordance with its domestic regulatory framework or Annex I; in calculating the time within which the Party of import is to respond, the number of days it has to wait for additional relevant information shall not be taken into account; or
(d) Informing the notifier that the period specified in this paragraph is extended by a defined period of time.

4. Except in a case in which consent is unconditional, a decision under paragraph 3 above, shall set out the reasons on which it is based.
5. A failure by the Party of import to communicate its decision within two hundred and seventy days of the date of receipt of notification shall not imply its consent to an intentional transboundary movement.
6. Lack of scientific certainty due to insufficient relevant scientific information and knowledge regarding the extent of the potential adverse effects of a living modified organism on the conservation and sustainable use of biological diversity in the Party of import, taking also into account risks to human health, shall not prevent that Party from taking a decision, as appropriate, with regard to the import of the living modified organism in question as referred to in paragraph 3 above, in order to avoid or minimize such potential adverse effects.
7. The Conference of the Parties serving as the meeting of the Parties shall, at its first meeting, decide upon appropriate procedures and mechanisms to facilitate decision-making by Parties of import.

Article 11

PROCEDURE FOR LIVING MODIFIED ORGANISMS INTENDED FOR DIRECT USE AS FOOD OR FEED, OR FOR PROCESSING

1. A Party that makes a final decision regarding domestic use, including placing on the market, of a living modified organism that may be subject to transboundary movement for direct use as food or feed, or for processing shall, within fifteen days of making that decision, inform the Parties through the Biosafety Clearing-House. This information shall contain, at a minimum, the information specified in Annex II. The Party shall provide a copy of the information, in writing, to the national focal point of each Party that informs the Secretariat in advance that it does not have access to the Biosafety Clearing-House. This provision shall not apply to decisions regarding field trials.
2. The Party making a decision under paragraph 1 above, shall ensure that there is a legal requirement for the accuracy of information provided by the applicant.

3. Any Party may request additional information from the authority identified in paragraph (b) of Annex II.
4. A Party may take a decision on the import of living modified organisms intended for direct use as food or feed, or for processing, under its domestic regulatory framework that is consistent with the objective of this Protocol.
5. Each Party shall make available to the Biosafety Clearing-House copies of any national laws, regulations and guidelines applicable to the import of living modified organisms intended for direct use as food or feed, or for processing, if available.
6. A developing country Party or a Party with an economy in transition may, in the absence of the domestic regulatory framework referred to in paragraph 4 above, and in exercise of its domestic jurisdiction, declare through the Biosafety Clearing-House that its decision prior to the first import of a living modified organism intended for direct use as food or feed, or for processing, on which information has been provided under paragraph 1 above, will be taken according to the following:
   (a) A risk assessment undertaken in accordance with Annex III; and
   (b) A decision made within a predictable timeframe, not exceeding two hundred and seventy days.
7. Failure by a Party to communicate its decision according to paragraph 6 above, shall not imply its consent or refusal to the import of a living modified organism intended for direct use as food or feed, or for processing, unless otherwise specified by the Party.
8. Lack of scientific certainty due to insufficient relevant scientific information and knowledge regarding the extent of the potential adverse effects of a living modified organism on the conservation and sustainable use of biological diversity in the Party of import, taking also into account risks to human health, shall not prevent that Party from taking a decision, as appropriate, with regard to the import of that living modified organism intended for direct use as food or feed, or for processing, in order to avoid or minimize such potential adverse effects.
9. A Party may indicate its needs for financial and technical assistance and capacity-building with respect to living modified organisms intended for direct use as food or feed, or for processing. Parties shall cooperate to meet these needs in accordance with Articles 22 and 28.

Article 12
REVIEW OF DECISIONS

1. A Party of import may, at any time, in light of new scientific information on potential adverse effects on the conservation and sustainable use of biological diversity, taking also into account risks to human health, review and change a decision regarding an intentional transboundary movement. In such case, the Party shall, within thirty days, inform any notifier that has previously notified movements of the living modified organism

referred to in such decision, as well as the Biosafety Clearing-House, and shall set out the reasons for its decision.

2. A Party of export or a notifier may request the Party of import to review a decision it has made in respect of it under Article 10 where the Party of export or the notifier considers that:
   (a) A change in circumstances has occurred that may influence the outcome of the risk assessment upon which the decision was based; or
   (b) Additional relevant scientific or technical information has become available.
3. The Party of import shall respond in writing to such a request within ninety days and set out the reasons for its decision.
4. The Party of import may, at its discretion, require a risk assessment for subsequent imports.

Article 13
SIMPLIFIED PROCEDURE

1. A Party of import may, provided that adequate measures are applied to ensure the safe intentional transboundary movement of living modified organisms in accordance with the objective of this Protocol, specify in advance to the Biosafety Clearing-House:
   (a) Cases in which intentional transboundary movement to it may take place at the same time as the movement is notified to the Party of import; and
   (b) Imports of living modified organisms to it to be exempted from the advance informed agreement procedure.

   Notifications under subparagraph (a) above, may apply to subsequent similar movements to the same Party.
2. The information relating to an intentional transboundary movement that is to be provided in the notifications referred to in paragraph 1 (a) above, shall be the information specified in Annex I.

Article 14
BILATERAL, REGIONAL AND MULTILATERAL AGREEMENTS AND ARRANGEMENTS

1. Parties may enter into bilateral, regional and multilateral agreements and arrangements regarding intentional transboundary movements of living modified organisms, consistent with the objective of this Protocol and provided that such agreements and arrangements do not result in a lower level of protection than that provided for by the Protocol.
2. The Parties shall inform each other, through the Biosafety Clearing-House, of any such bilateral, regional and multilateral agreements and arrangements that they have entered into before or after the date of entry into force of this Protocol.

3. The provisions of this Protocol shall not affect intentional transboundary movements that take place pursuant to such agreements and arrangements as between the parties to those agreements or arrangements.
4. Any Party may determine that its domestic regulations shall apply with respect to specific imports to it and shall notify the Biosafety Clearing-House of its decision.

Article 15
RISK ASSESSMENT

1. Risk assessments undertaken pursuant to this Protocol shall be carried out in a scientifically sound manner, in accordance with Annex III and taking into account recognized risk assessment techniques. Such risk assessments shall be based, at a minimum, on information provided in accordance with Article 8 and other available scientific evidence in order to identify and evaluate the possible adverse effects of living modified organisms on the conservation and sustainable use of biological diversity, taking also into account risks to human health.
2. The Party of import shall ensure that risk assessments are carried out for decisions taken under Article 10. It may require the exporter to carry out the risk assessment.
3. The cost of risk assessment shall be borne by the notifier if the Party of import so requires.

Article 16
RISK MANAGEMENT

1. The Parties shall, taking into account Article 8 (g) of the Convention, establish and maintain appropriate mechanisms, measures and strategies to regulate, manage and control risks identified in the risk assessment provisions of this Protocol associated with the use, handling and transboundary movement of living modified organisms.
2. Measures based on risk assessment shall be imposed to the extent necessary to prevent adverse effects of the living modified organism on the conservation and sustainable use of biological diversity, taking also into account risks to human health, within the territory of the Party of import.
3. Each Party shall take appropriate measures to prevent unintentional transboundary movements of living modified organisms, including such measures as requiring a risk assessment to be carried out prior to the first release of a living modified organism.
4. Without prejudice to paragraph 2 above, each Party shall endeavour to ensure that any living modified organism, whether imported or locally developed, has undergone an appropriate period of observation that is commensurate with its life-cycle or generation time before it is put to its intended use.

5. Parties shall cooperate with a view to:
   (a) Identifying living modified organisms or specific traits of living modified organisms that may have adverse effects on the conservation and sustainable use of biological diversity, taking also into account risks to human health; and
   (b) Taking appropriate measures regarding the treatment of such living modified organisms or specific traits.

Article 17
UNINTENTIONAL TRANSBOUNDARY MOVEMENTS AND EMERGENCY MEASURES

1. Each Party shall take appropriate measures to notify affected or potentially affected States, the Biosafety Clearing-House and, where appropriate, relevant international organizations, when it knows of an occurrence under its jurisdiction resulting in a release that leads, or may lead, to an unintentional transboundary movement of a living modified organism that is likely to have significant adverse effects on the conservation and sustainable use of biological diversity, taking also into account risks to human health in such States. The notification shall be provided as soon as the Party knows of the above situation.
2. Each Party shall, no later than the date of entry into force of this Protocol for it, make available to the Biosafety Clearing-House the relevant details setting out its point of contact for the purposes of receiving notifications under this Article.
3. Any notification arising from paragraph 1 above, should include:
   (a) Available relevant information on the estimated quantities and relevant characteristics and/or traits of the living modified organism;
   (b) Information on the circumstances and estimated date of the release, and on the use of the living modified organism in the originating Party;
   (c) Any available information about the possible adverse effects on the conservation and sustainable use of biological diversity, taking also into account risks to human health, as well as available information about possible risk management measures;
   (d) Any other relevant information; and
   (e) A point of contact for further information.
4. In order to minimize any significant adverse effects on the conservation and sustainable use of biological diversity, taking also into account risks to human health, each Party, under whose jurisdiction the release of the living modified organism referred to in paragraph 1 above, occurs, shall immediately consult the affected or potentially affected States to enable them to determine appropriate responses and initiate necessary action, including emergency measures.

Article 18
HANDLING, TRANSPORT, PACKAGING AND IDENTIFICATION

1. In order to avoid adverse effects on the conservation and sustainable use of biological diversity, taking also into account risks to human health, each Party shall take necessary measures to require that living modified organisms that are subject to intentional transboundary movement within the scope of this Protocol are handled, packaged and transported under conditions of safety, taking into consideration relevant international rules and standards.
2. Each Party shall take measures to require that documentation accompanying:
   (a) Living modified organisms that are intended for direct use as food or feed, or for processing, clearly identifies that they "may contain" living modified organisms and are not intended for intentional introduction into the environment, as well as a contact point for further information. The Conference of the Parties serving as the meeting of the Parties to this Protocol shall take a decision on the detailed requirements for this purpose, including specification of their identify and any unique identification, no later than two years after the date of entry into force of this Protocol;
   (b) Living modified organisms that are destined for contained use clearly identifies them as living modified organisms; and specifies any requirements for safe handling, storage, transport and use, the contact point for further information, including the name and address of the individual and institution to whom the living modified organisms are consigned; and
   (c) Living modified organisms that are intended for intentional introduction into the environment of the Party of import and any other living modified organisms within the scope of the Protocol, clearly identifies them as living modified organisms; specifies the identity and relevant traits and/or characteristics, any requirements for the safe handling, storage, transport and use, the contact point for further information and, as appropriate, the name and address of the importer and exporter; and contains a declaration that the movement is in conformity with the requirements of this Protocol applicable to the exporter.
3. The Conference of the Parties serving as the meeting of the Parties to this Protocol shall consider the need for and modalities of developing standards with regard to identification, handling, packaging and transport practices, in consultation with other relevant international bodies.

Article 19
COMPETENT NATIONAL AUTHORITIES AND NATIONAL FOCAL POINTS

1. Each Party shall designate one national focal point to be responsible on its behalf for liaison with the Secretariat. Each Party shall also designate

one or more competent national authorities, which shall be responsible for performing the administrative functions required by this Protocol and which shall be authorized to act on its behalf with respect to those functions. A Party may designate a single entity to fulfill the functions of both focal point and competent national authority.

2. Each Party shall, no later than the date of entry into force of this Protocol for it, notify the Secretariat of the names and addresses of its focal point and its competent national authority or authorities. Where a Party designates more than one competent national authority, it shall convey to the Secretariat, with its notification thereof, relevant information on the respective responsibilities of those authorities. Where applicable, such information shall, at a minimum, specify which competent authority is responsible for which type of living modified organism. Each Party shall forthwith notify the Secretariat of any changes in the designation of its national focal point or in the name and address of responsibilities of its competent national authority or authorities.
3. The Secretariat shall forthwith inform the Parties of the notifications it receives under paragraph 2 above, and shall also make such information available through the Biosafety Clearing-House.

Article 20
INFORMATION SHARING AND THE BIOSAFETY CLEARING-HOUSE

1. A Biosafety Clearing-House is hereby established as part of the clearing-house mechanism under Article 18, paragraph 3, of the Convention, in order to:
   (a) Facilitate the exchange of scientific, technical, environmental and legal information on, and experience with, living modified organisms; and
   (b) Assist Parties to implement the Protocol, taking into account the special needs of developing country Parties, in particular the least developed and small island developing States among them, and countries with economies in transition as well as countries that are centres of origin and centres of genetic diversity.
2. The Biosafety Clearing-House shall serve as a means through which information is made available for the purposes of paragraph 1 above. It shall provide access to information made available by the Parties relevant to the implementation of the Protocol. It shall also provide access, where possible, to other international biosafety information exchange mechanisms.
3. Without prejudice to the protection of confidential information, each Party shall make available to the Biosafety Clearing-House any information required to be made available to the Biosafety Clearing-House under this Protocol, and:

(a) Any existing laws, regulations and guidelines for implementation of the Protocol, as well as information required by the Parties for the advance informed agreement procedure;
(b) Any bilateral, regional and multilateral agreements and arrangements;
(c) Summaries of its risk assessments or environmental reviews of living modified organisms generated by its regulatory process, and carried out in accordance with Article 15, including, where appropriate, relevant information regarding products thereof, namely, processed materials that are of living modified organism origin, containing detectable novel combinations of replicable genetic material obtained through the use of modern biotechnology;
(d) Its final decisions regarding the importation or release of living modified organisms; and
(e) Reports submitted by it pursuant to Article 33, including those on implementation of the advance informed agreement procedure.

4. The modalities of the operation of the Biosafety Clearing-House, including reports on its activities, shall be considered and decided upon by the Conference of the Parties serving as the meeting of the Parties to this Protocol at its first meeting, and kept under review thereafter.

Article 21
CONFIDENTIAL INFORMATION

1. The Party of import shall permit the notifier to identify information submitted under the procedures of this Protocol or required by the Party of import as part of the advance informed agreement procedure of the Protocol that is to be treated as confidential. Justification shall be given in such cases upon request.
2. The Party of import shall consult the notifier if it decides that information identified by the notifier as confidential does not qualify for such treatment and shall, prior to any disclosure, inform the notifier of its decision, providing reasons on request, as well as an opportunity for consultation and for an internal review of the decision prior to disclosure.
3. Each Party shall protect confidential information received under this Protocol, including any confidential information received in the context of the advance informed agreement procedure of the Protocol. Each Party shall ensure that it has procedures to protect such information and shall protect the confidentiality of such information in a manner no less favourable than its treatment of confidential information in connection with domestically produced living modified organisms.
4. The Party of import shall not use such information for a commercial purpose, except with the written consent of the notifier.
5. If a notifier withdraws or has withdrawn a notification, the Party of import shall respect the confidentiality of commercial and industrial information, including research and development information as well as information on which the Party and the notifier disagree as to its confidentiality.

6. Without prejudice to paragraph 5 above, the following information shall not be considered confidential:
   (a) The name and address of the notifier;
   (b) A general description of the living modified organism or organisms;
   (c) A summary of the risk assessment of the effects on the conservation and sustainable use of biological diversity, taking also into account risks to human health; and
   (d) Any methods and plans for emergency response.

Article 22
CAPACITY-BUILDING

1. The Parties shall cooperate in the development and/or strengthening of human resources and institutional capacities in biosafety, including biotechnology to the extent that it is required for biosafety, for the purpose of the effective implementation of this Protocol, in developing country Parties, in particular the least developed and small island developing States among them, and in Parties with economies in transition, including through existing global, regional, subregional and national institutions and organizations and, as appropriate, through facilitating private sector involvement.
2. For the purposes of implementing paragraph 1 above, in relation to cooperation, the needs of developing country Parties, in particular the least developed and small island developing States among them, for financial resources and access to and transfer of technology and know-how in accordance with the relevant provisions of the Convention, shall be taken fully into account for capacity-building in biosafety. Cooperation in capacity-building shall, subject to the different situation, capabilities and requirements of each Party, include scientific and technical training in the proper and safe management of biotechnology, and in the use of risk assessment and risk management for biosafety, and the enhancement of technological and institutional capacities in biosafety. The needs of Parties with economies in transition shall also be taken fully into account for such capacity-building in biosafety.

Article 23
PUBLIC AWARENESS AND PARTICIPATION

1. The Parties shall:
   (a) Promote and facilitate public awareness, education and participation concerning the safe transfer, handling and use of living modified organisms in relation to the conservation and sustainable use of biological diversity, taking also into account risks to human health. In doing so, the Parties shall cooperate, as appropriate, with other States and international bodies;

(b) Endeavour to ensure that public awareness and education encompass access to information on living modified organisms identified in accordance with this Protocol that may be imported.

2. The Parties shall, in accordance with their respective laws and regulations, consult the public in the decision-making process regarding living modified organisms and shall make the results of such decisions available to the public, while respecting confidential information in accordance with Article 21.
3. Each Party shall endeavour to inform its public about the means of public access to the Biosafety Clearing-House.

Article 24
NON-PARTIES

1. Transboundary movements of living modified organisms between Parties and non-Parties shall be consistent with the objective of this Protocol. The Parties may enter into bilateral, regional and multilateral agreements and arrangements with non-Parties regarding such transboundary movements.
2. The Parties shall encourage non-Parties to adhere to this Protocol and to contribute appropriate information to the Biosafety Clearing-House on living modified organisms released in, or moved into or out of, areas within their national jurisdictions.

Article 25
ILLEGAL TRANSBOUNDARY MOVEMENTS

1. Each Party shall adopt appropriate domestic measures aimed at preventing and, if appropriate, penalizing transboundary movements of living modified organisms carried out in contravention of its domestic measures to implement this Protocol. Such movements shall be deemed illegal transboundary movements.
2. In the case of an illegal transboundary movement, the affected Party may request the Party of origin to dispose, at its own expense, of the living modified organism in question by repatriation or destruction, as appropriate.
3. Each Party shall make available to the Biosafety Clearing-House information concerning cases of illegal transboundary movements pertaining to it.

Article 26
SOCIO-ECONOMIC CONSIDERATIONS

1. The Parties, in reaching a decision on import under this Protocol or under its domestic measures implementing the Protocol, may take into account, consistent with their international obligations, socio-economic

considerations arising from the impact of living modified organisms on the conservation and sustainable use of biological diversity, especially with regard to the value of biological diversity to indigenous and local communities.

2. The Parties are encouraged to cooperate on research and information exchange on any socio-economic impacts of living modified organisms, especially on indigenous and local communities.

Article 27
LIABILITY AND REDRESS

The Conference of the Parties serving as the meeting of the Parties to this Protocol shall, at its first meeting, adopt a process with respect to the appropriate elaboration of international rules and procedures in the field of liability and redress for damage resulting from transboundary movements of living modified organisms, analysing and taking due account of the ongoing processes in international law on these matters, and shall endeavour to complete this process within four years.

Article 28
FINANCIAL MECHANISM AND RESOURCES

1. In considering financial resources for the implementation of this Protocol, the Parties shall take into account the provisions of Article 20 of the Convention.
2. The financial mechanism established in Article 21 of the Convention shall, through the institutional structure entrusted with its operation, be the financial mechanism for this Protocol.
3. Regarding the capacity-building referred to in Article 22 of this Protocol, the Conference of the Parties serving as the meeting of the Parties to this Protocol, in providing guidance with respect to the financial mechanism referred to in paragraph 2 above, for consideration by the Conference of the Parties, shall take into account the need for financial resources by developing country Parties, in particular the least developed and the small island developing States among them.
4. In the context of paragraph 1 above, the Parties shall also take into account the needs of the developing country Parties, in particular the least developed and the small island developing States among them, and of the Parties with economies in transition, in their efforts to identify and implement their capacity-building requirements for the purposes of the implementation of this Protocol.
5. The guidance to the financial mechanism of the Convention in relevant decisions of the Conference of the Parties, including those agreed before the adoption of this Protocol, shall apply, *mutatis mutandis,* to the provisions of this Article.

6. The developed country Parties may also provide, and the developing country Parties and the Parties with economies in transition avail themselves of, financial and technological resources for the implementation of the provisions of this Protocol through bilateral, regional and multilateral channels.

Article 29

CONFERENCE OF THE PARTIES SERVING AS THE MEETING OF THE PARTIES TO THIS PROTOCOL

1. The Conference of the Parties shall serve as the meeting of the Parties to this Protocol.
2. Parties to the Convention that are not Parties to this Protocol may participate as observers in the proceedings of any meeting of the Conference of the Parties serving as the meeting of the Parties to this Protocol. When the Conference of the Parties serves as the meeting of the Parties to this Protocol, decisions under this Protocol shall be taken only by those that are Parties to it.
3. When the Conference of the Parties serves as the meeting of the Parties to this Protocol, any member of the bureau of the Conference of the Parties representing a Party to the Convention but, at that time, not a Party to this Protocol, shall be substituted by a member to be elected by and from among the Parties to this Protocol.
4. The Conference of the Parties serving as the meeting of the Parties to this Protocol shall keep under regular review the implementation of this Protocol and shall make, within its mandate, the decisions necessary to promote its effective implementation. It shall perform the functions assigned to it by this Protocol and shall:
   (a) Make recommendations on any matters necessary for the implementation of this Protocol;
   (b) Establish such subsidiary bodies as are deemed necessary for the implementation of this Protocol;
   (c) Seek and utilize, where appropriate, the services and cooperation of, and information provided by, competent international organizations and intergovernmental and non-governmental bodies;
   (d) Establish the form and the intervals for transmitting the information to be submitted in accordance with Article 33 of this Protocol and consider such information as well as reports submitted by any subsidiary body;
   (e) Consider and adopt, as required, amendments to this Protocol and its annexes, as well as any additional annexes to this Protocol, that are deemed necessary for the implementation of this Protocol; and
   (f) Exercise such other functions as may be required for the implementation of this Protocol.

5. The rules of procedure of the Conference of the Parties and financial rules of the Convention shall be applied, *mutatis mutandis,* under this Protocol, except as may be otherwise decided by consensus by the Conference of the Parties serving as the meeting of the Parties to this Protocol.
6. The first meeting of the Conference of the Parties serving as the meeting of the Parties to this Protocol shall be convened by the Secretariat in conjunction with the first meeting of the Conference of the Parties that is scheduled after the date of the entry into force of this Protocol. Subsequent ordinary meetings of the Conference of the Parties serving as the meeting of the Parties to this Protocol shall be held in conjunction with ordinary meetings of the Conference of the Parties, unless otherwise decided by the Conference of the Parties serving as the meeting of the Parties to this Protocol.
7. Extraordinary meetings of the Conference of the Parties serving as the meeting of the Parties to this Protocol shall be held at such other times as may be deemed necessary by the Conference of the Parties serving as the meeting of the Parties to this Protocol, or at the written request of any Party, provided that, within six months of the request being communicated to the Parties by the Secretariat, it is supported by at least one third of the Parties.
8. The United Nations, its specialized agencies and the International Atomic Energy Agency, as well as any State member thereof or observers thereto not party to the Convention, may be represented as observers at meetings of the Conference of the Parties serving as the meeting of the Parties to this Protocol. Any body or agency, whether national or international, governmental or non-governmental, that is qualified in matters covered by this Protocol and that has informed the Secretariat of its wish to be represented at a meeting of the Conference of the Parties serving as a meeting of the Parties to this Protocol as an observer, may be so admitted, unless at least one third of the Parties present object. Except as otherwise provided in this Article, the admission and participation of observers shall be subject to the rules of procedure, as referred to in paragraph 5 above.

Article 30
SUBSIDIARY BODIES

1. Any subsidiary body established by or under the Convention may, upon a decision by the Conference of the Parties serving as the meeting of the Parties to this Protocol, serve the Protocol, in which case the meeting of the Parties shall specify which functions that body shall exercise.
2. Parties to the Convention that are not Parties to this Protocol may participate as observers in the proceedings of any meeting of any such subsidiary bodies. When a subsidiary body of the Convention serves as a subsidiary body to this Protocol, decisions under the Protocol shall be taken only by the Parties to the Protocol.

3. When a subsidiary body of the Convention exercises its functions with regard to matters concerning this Protocol, any member of the bureau of that subsidiary body representing a Party to the Convention but, at that time, not a Party to the Protocol, shall be substituted by a member to be elected by and from among the Parties to the Protocol.

Article 31
SECRETARIAT

1. The Secretariat established by Article 24 of the Convention shall serve as the secretariat to this Protocol.
2. Article 24, paragraph 1, of the Convention on the functions of the Secretariat shall apply, *mutatis mutandis,* to this Protocol.
3. To the extent that they are distinct, the costs of the secretariat services for this Protocol shall be met by the Parties hereto. The Conference of the Parties serving as the meeting of the Parties to this Protocol shall, at its first meeting, decide on the necessary budgetary arrangements to this end.

Article 32
RELATIONSHIP WITH THE CONVENTION

Except as otherwise provided in this Protocol, the provisions of the Convention relating to its Protocols shall apply to this Protocol.

Article 33
MONITORING AND REPORTING

Each Party shall monitor the implementation of its obligations under this Protocol, and shall, at intervals to be determined by the Conference of the Parties serving as the meeting of the Parties to this Protocol, report to the Conference of the Parties serving as the meeting of the Parties to this Protocol on measures that it has taken to implement the Protocol.

Article 34
COMPLIANCE

The Conference of the Parties serving as the meeting of the Parties to this Protocol shall, at its first meeting, consider and approve cooperative procedures and institutional mechanisms to promote compliance with the provisions of this Protocol and to address cases of non-compliance. These procedures and mechanisms shall include provisions to offer advice or assistance, where appropriate. They shall be separate from, and without prejudice to, the dispute settlement procedures and mechanisms established by Article 27 of the Convention.

Article 35
ASSESSMENT AND REVIEW

The Conference of the Parties serving as the meeting of the Parties to this Protocol shall undertake, five years after the entry into force of this Protocol and at least every five years thereafter, an evaluation of the effectiveness of the Protocol, including an assessment of its procedures and annexes.

Article 36
SIGNATURE

This Protocol shall be open for signature at the United Nations Office at Nairobi by States and regional economic integration organizations from 15 to 26 May 2000, and at United Nations Headquarters in New York from 5 June 2000 to 4 June 2001.

Article 37
ENTRY INTO FORCE

1. This Protocol shall enter into force on the ninetieth day after the date of deposit of the fiftieth instrument of ratification, acceptance, approval or accession by States or regional economic integration organizations that are Parties to the Convention.
2. This Protocol shall enter into force for a State or regional economic integration organization that ratifies, accepts or approves this Protocol or accedes thereto after its entry into force pursuant to paragraph 1 above, on the ninetieth day after the date on which that State or regional economic integration organization deposits its instrument of ratification, acceptance, approval or accession, or on the date on which the Convention enters into force for that State or regional economic integration organization, whichever shall be the later.
3. For the purposes of paragraphs 1 and 2 above, any instrument deposited by a regional economic integration organization shall not be counted as additional to those deposited by member States of such organization.

Article 38
RESERVATIONS

No reservations may be made to this Protocol.

Article 39
WITHDRAWAL

1. At any time after two years from the date on which this Protocol has entered into force for a Party, that Party may withdraw from the Protocol by giving written notification to the Depositary.

2. Any such withdrawal shall take place upon expiry of one year after the date of its receipt by the Depositary, or on such later date as may be specified in the notification of the withdrawal.

Article 40
AUTHENTIC TEXTS

The original of this Protocol, of which the Arabic, Chinese, English, French, Russian and Spanish texts are equally authentic, shall be deposited with the Secretary-General of the United Nations.

IN WITNESS THEREOF the undersigned, being duly authorized to that effect, have signed this Protocol.

DONE at Montreal on this twenty-ninth day of January, two thousand.

Annex I
INFORMATION REQUIRED IN NOTIFICATIONS UNDER ARTICLES 8, 10 AND 13

(a) Name, address and contact details of the exporter.
(b) Name, address and contact details of the importer.
(c) Name and identity of the living modified organism, as well as the domestic classification, if any, of the biosafety level of the living modified organism in the State of export.
(d) Intended date or dates of the transboundary movement, if known.
(e) Taxonomic status, common name, point of collection or acquisition, and characteristics of recipient organism or parental organisms related to biosafety.
(f) Centres of origin and centres of genetic diversity, if known, of the recipient organism and/or the parental organisms and a description of the habitats where the organisms may persist or proliferate.
(g) Taxonomic status, common name, point of collection or acquisition, and characteristics of the donor organism or organisms related to biosafety.
(h) Description of the nucleic acid or the modification introduced, the technique used, and the resulting characteristics of the living modified organism.
(i) Intended use of the living modified organism or products thereof, namely, processed materials that are of living modified organism origin, containing detectable novel combinations of replicable genetic material obtained through the use of modern biotechnology.
(j) Quantity or volume of the living modified organism to be transferred.
(k) A previous and existing risk assessment report consistent with *Annex III*.
(l) Suggested methods for the safe handling, storage, transport and use, including packaging, labelling, documentation, disposal and contingency procedures, where appropriate.

(m) Regulatory status of the living modified organism within the State of export (for example, whether it is prohibited in the State of export, whether there are other restrictions, or whether it has been approved for general release) and, if the living modified organism is banned in the State of export, the reason or reasons for the ban.
(n) Result and purpose of any notification by the exporter to other States regarding the living modified organism to be transferred.
(o) A declaration that the above-mentioned information is factually correct.

Annex II

INFORMATION REQUIRED CONCERNING LIVING MODIFIED ORGANISMS INTENDED FOR DIRECT USE AS FOOD OR FEED, OR FOR PROCESSING UNDER ARTICLE 11

(a) The name and contact details of the applicant for a decision for domestic use.
(b) The name and contact details of the authority responsible for the decision.
(c) Name and identity of the living modified organism.
(d) Description of the gene modification, the technique used, and the resulting characteristics of the living modified organism.
(e) Any unique identification of the living modified organism.
(f) Taxonomic status, common name, point of collection or acquisition, and characteristics of recipient organism or parental organisms related to biosafety.
(g) Centres of origin and centres of genetic diversity, if known, of the recipient organism and /or the parental organisms and a description of the habitats where the organisms may persist or proliferate.
(h) Taxonomic status, common name, point of collection or acquisition, and characteristics of the donor organism or organisms related to biosafety.
(i) Approved uses of the living modified organism.
(j) A risk assessment report consistent with *Annex III*.
(k) Suggested methods for the safe handling, storage, transport and use, including packaging, labelling, documentation, disposal and contingency procedures, where appropriate.

Annex III

RISK ASSESSMENT

Objective

1. The objective of risk assessment, under this Protocol, is to identify and evaluate the potential adverse effects of living modified organisms on the conservation and sustainable use of biological diversity in the likely potential receiving environment, taking also into account risks to human health.

Use of risk assessment

2. Risk assessment is, *inter alia*, used by competent authorities to make informed decisions regarding living modified organisms.

General principles

3. Risk assessment should be carried out in a scientifically sound and transparent manner, and can take into account expert advice of, and guidelines developed by, relevant international organizations.
4. Lack of scientific knowledge or scientific consensus should not necessarily be interpreted as indicating a particular level of risk, an absence of risk, or an acceptable risk.
5. Risks associated with living modified organisms or products thereof, namely, processed materials that are of living modified organism origin, containing detectable novel combinations of replicable genetic material obtained through the use of modern biotechnology, should be considered in the context of the risks posed by the non-modified recipients or parental organisms in the likely potential receiving environment.
6. Risk assessment should be carried out on a case-by-case basis. The required information may vary in nature and level of detail from case to case, depending on the living modified organism concerned, its intended use and the likely potential receiving environment.

Methodology

7. The process of risk assessment may on the one hand give rise to a need for further information about specific subjects, which may be identified and requested during the assessment process, while on the other hand information on other subjects may not be relevant in some instances.
8. To fulfill its objective, risk assessment entails, as appropriate, the following steps:
   (a) An identification of any novel genotypic and phenotypic characteristics associated with the living modified organism that may have adverse effects on biological diversity in the likely potential receiving environment, taking also into account risks to human health;
   (b) An evaluation of the likelihood of these adverse effects being realized, taking into account the level and kind of exposure of the likely potential receiving environment to the living modified organism;
   (c) An evaluation of the consequences should these adverse effects be realized;
   (d) An estimation of the overall risk posed by the living modified organism based on the evaluation of the likelihood and consequences of the identified adverse effects being realized;
   (e) A recommendation as to whether or not the risks are acceptable or manageable, including, where necessary, identification of strategies to manage these risks; and

(f) Where there is uncertainty regarding the level of risk, it may be addressed by requesting further information on the specific issues of concern or by implementing appropriate risk management strategies and/or monitoring the living modified organism in the receiving environment.

Points to consider

9. Depending on the case, risk assessment takes into account the relevant technical and scientific details regarding the characteristics of the following subjects:
   (a) *Recipient organism or parental organisms.* The biological characteristics of the recipient organism or parental organisms, including information on taxonomic status, common name, origin, centres of origin and centres of genetic diversity, if known, and a description of the habitat where the organisms may persist or proliferate;
   (b) *Donor organism or organisms.* Taxonomic status and common name, source, and the relevant biological characteristics of the donor organisms;
   (c) *Vector.* Characteristics of the vector, including its identity, if any, and its source or origin, and its host range;
   (d) *Insert or inserts and/or characteristics of modification.* Genetic characteristics of the inserted nucleic acid and the function it specifies, and/or characteristics of the modification introduced;
   (e) *Living modified organism.* Identity of the living modified organism, and the differences between the biological characteristics of the living modified organism and those of the recipient organism or parental organisms;
   (f) *Detection and identification of the living modified organism.* Suggested detection and identification methods and their specificity, sensitivity and reliability;
   (g) *Information relating to the intended use.* Information relating to the intended use of the living modified organism, including new or changed use compared to the recipient organism or parental organisms; and
   (h) *Receiving environment.* Information on the location, geographical, climatic and ecological characteristics, including relevant information on biological diversity and centres of origin of the likely potential receiving environment.

Legislation is in effect in the European Union that requires labeling of genetically engineered foods. The following document is the legislation as passed on May 26, 1998.

L 159/4 EN Official Journal of the European Communities 3. 6. 98

COUNCIL REGULATION (EC) No 1139/98

of 26 May 1998

concerning the compulsory indication of the labelling of certain foodstuffs produced from genetically modified organisms of particulars other than those provided for in Directive 79/112/EEC

THE COUNCIL OF THE EUROPEAN UNION,

Having regard to the Treaty establishing the European Community,

Having regard to Council Directive 79/112/EEC of 18 December 1978 on the approximation of the laws of the Member States relating to the labelling, presentation and advertising of foodstuffs ([1]), and in particular Article 4(2) thereof,

Having regard to the proposal from the Commission,

(1) Whereas, in accordance with the provisions of Part C of Council Directive 90/220/EEC of 23 April 1990 on the deliberate release into the environment of genetically modified organisms ([2]), consents have been given for the placing on the market of certain genetically modified products by Commission Decision 96/281/EC of 3 April 1996 concerning the placing on the market of genetically modified soya beans (*Glycine max* L.) with increased tolerance to the herbicide glyphosate, pursuant to Council Directive 90/220/EEC ([3]), and by Commission Decision 97/98/EC of 23 January 1997 concerning the placing on the market of genetically modified maize (*Zea mays* L.) with the combined modification for insecticidal properties conferred by the Bt-endotoxin gene and increased tolerance to the herbicide glufosinate ammonium pursuant to Council Directive 90/220/EEC ([4]);

(2) Whereas in accordance with Directive 90/220/EEC there were no safety grounds for mentioning on the label of genetically modified soya beans (*Glycine max* L.) or of genetically modified maize (*Zea mays* L). that they were obtained by genetic modification techniques;

(3) Whereas Directive 90/220/EEC does not cover non-viable products derived from genetically modified organisms (hereinafter referred to as 'GMOs');

(4) Whereas certain Member States have taken measures in respect of the labelling of foods and food ingredients produced from the products concerned; whereas differences between those measures are liable to impede the free movement of those foods and food ingredients and thereby adversely affect the functioning of the internal market; whereas it is therefore necessary to adopt uniform Community labelling rules for the products concerned;

(5) Whereas Regulation (EC) No 258/97 of the European Parliament and of the Council of 27 January 1997 concerning novel foods and novel food ingredients ([5]), lays down, in Article 8, additional specific labelling requirements in order to ensure proper information for the final consumer; whereas those additional specific labelling requirements do not apply to foods or food ingredients which were used for human consumption to a significant degree within the Community before the entry into force of Regulation (EC) 258/97 and are for that reason considered not to be novel;

(6) Whereas, in order to prevent distortions of competition, labelling rules for the information of the final consumer based on the same principles should apply to foods and food ingredients consisting of or derived from GMOs which were placed on the market before the entry into force of Regulation (EC) No 258/97 pursuant to a consent given under Directive 90/220/EEC, and to foods and food ingredients which are placed on the market thereafter;

(7) Whereas, therefore, Commission Regulation (EC) No 1813/97 of 19 September 1997 concerning the compulsory indication on the labelling of certain foodstuffs produced from genetically modified organisms of particulars other than those provided for in Directive 79/112/EEC ([6]) laid down general labelling rules for the abovementioned products;

(8) Whereas it is now urgent to lay down detailed uniform Community rules for the labelling of the foodstuffs covered by Regulation (EC) No 1813/97;

([1]) OJ L 33, 8. 2. 1979, p. 1. Directive as last amended by Directive 97/4/EC of the European Parliament and of the Council (OJ L 43, 14. 2. 1997, p. 21).
([2]) OJ L 117, 8. 5. 1990, p. 15. Directive as last amended by Directive 97/35/EC (OJ L 169, 27. 6. 1997, p. 72).
([3]) OJ L 107, 30. 4. 1996, p. 10.
([4]) OJ L 31, 1. 2. 1997, p. 69.
([5]) OJ L 43, 14. 2. 1997, p. 1.
([6]) OJ L 257, 20. 9. 1997, p. 7.

Section VIII

Web Sites

There is perhaps no issue whose politics have been driven so unremittingly by the Internet as bioengineering of food. Perhaps because so much of the impetus behind the protest against GMOs has originated in foreign countries, or perhaps because of the vast number and diversity of interests involved, or perhaps because cyber activism and the genetic modification of food came of age at virtually the same time — whatever the reason, this particular political war is a product of the electronic revolution.

In addition to the Web sites of the national environmental advocacy organizations, e.g., Greenpeace, Friends of the Earth, Environmental Defense, the Sierra Club, and those of the U.S. government agencies with regulatory responsibilities over GMOs — the Department of Agriculture, the EPA, the U.S. Food and Drug Administration — and foreign authorities as wide ranging as the U.K. Parliament, the European Union, the United Nations, and the World Health Organization, the following each have Web sites and links dedicated to GMOs.

Advocacy Groups like the Alliance for Better Foods, the Campaign for Food Safety, Consumer Alert, the Food Biotech Communications Network, Food First: Institute for Food and Development Policy, the Food Marketing Institute, the European Food Information Council, the International Center for Technology Assessment, and the Organic Consumers Association

Trade Associations like the American Seed Trade Association, the Canola Council of Canada, the National Cotton Industry, the National Food Processors Association, the Grocery Manufacturers of America, and the American Soybean Association

Farmers Organizations like the American Corn Growers Association, the American Farm Bureau Federation, the National Family Farm Coalition, and the National Farmers Union

Grain Processors like Archer Daniels Midland, ConAgra, the Corn Refiners Association, A.E. Staley, and the National Grain and Feed Association

Food and Agriculture Biotech Companies like Monsanto, Unilver, Novartis, DuPont, Nestle, Heinz, and AztraZeneca

Research Foundations like the Rockefeller Foundation, AgBiotechNet, Consumers Union, the National Academy of Sciences, the National Center for Food and Agricultural Policy, the National Center for Food Safety and Technology, the Nuffield Foundation, and the European Federation of Biotech

Issues

As noted, the international controversy over genetically modified food has been fomented and expanded largely in cyberspace. Whether or not public policy will be more enlightened and socially responsible for it remains to be seen. No doubt, the Internet has made available unprecedented levels of studies and information that reach only a tiny number of people by other means. More important, it has leveled the playing field for less well-funded, more obscure, and politically weaker interests, enabling them to carry out a successful guerrilla war against the powerful self-interested transnational corporations like Monsanto and Novartis as well as the many trade, biotech, and agricultural groups whose public relations campaigns in favor of genetically modified foods would otherwise dominate the policy arena.

On the other hand, the Internet is an open forum, and any manner of person, interest, or group has access to it. It thus becomes a tool for irresponsible as well as responsible parties, and for circulating inaccurate as well as accurate information. Its potential for use and abuse with respect to an issue like the bioengineering of food is thus rife with possibilities. Opponents of genetically modified foods have exploited the scientific uncertainty of their ultimate effects on human health and the ecosystem through an international network of communication. No other medium can spread fear and rumor as quickly and as pervasively as the Internet.

The virtues and limitations of the Internet as a political tool in the biotech wars are, of course, the same as they are generally, except that the novel and technologically sophisticated character of genetically modified foods exaggerate both. With these thoughts in mind, an examination of a range of Web sites would no doubt prove interesting and rewarding, not only as a study of political efficacy, but, just possibly, as a source of education and enlightenment on the subject itself.